SpringerBriefs in Applied Sciences and Technology

Computational Intelligence

Series Editor

Janusz Kacprzyk, Systems Research Institute, Polish Academy of Sciences, Warsaw, Poland

SpringerBriefs in Computational Intelligence are a series of slim high-quality publications encompassing the entire spectrum of Computational Intelligence. Featuring compact volumes of 50 to 125 pages (approximately 20,000-45,000 words), Briefs are shorter than a conventional book but longer than a journal article. Thus Briefs serve as timely, concise tools for students, researchers, and professionals.

Patricia Melin · Oscar Castillo

Type-3 Fuzzy Logic and Fractal Theory for Medical Diagnosis

Patricia Melin
Division of Graduate Studies
Tijuana Institute of Technology, TecNM
Tijuana, Baja California, Mexico

Oscar Castillo
Division of Graduate Studies
Tijuana Institute of Technology, TecNM
Tijuana, Baja California, Mexico

ISSN 2191-530X   ISSN 2191-5318 (electronic)
SpringerBriefs in Applied Sciences and Technology
ISSN 2625-3704   ISSN 2625-3712 (electronic)
SpringerBriefs in Computational Intelligence
ISBN 978-3-031-81654-3   ISBN 978-3-031-81655-0 (eBook)
https://doi.org/10.1007/978-3-031-81655-0

© The Editor(s) (if applicable) and The Author(s), under exclusive license to Springer Nature Switzerland AG 2025

This work is subject to copyright. All rights are solely and exclusively licensed by the Publisher, whether the whole or part of the material is concerned, specifically the rights of translation, reprinting, reuse of illustrations, recitation, broadcasting, reproduction on microfilms or in any other physical way, and transmission or information storage and retrieval, electronic adaptation, computer software, or by similar or dissimilar methodology now known or hereafter developed.
The use of general descriptive names, registered names, trademarks, service marks, etc. in this publication does not imply, even in the absence of a specific statement, that such names are exempt from the relevant protective laws and regulations and therefore free for general use.
The publisher, the authors and the editors are safe to assume that the advice and information in this book are believed to be true and accurate at the date of publication. Neither the publisher nor the authors or the editors give a warranty, expressed or implied, with respect to the material contained herein or for any errors or omissions that may have been made. The publisher remains neutral with regard to jurisdictional claims in published maps and institutional affiliations.

This Springer imprint is published by the registered company Springer Nature Switzerland AG
The registered company address is: Gewerbestrasse 11, 6330 Cham, Switzerland

If disposing of this product, please recycle the paper.

Preface

This book focuses on the utilization of type-3 fuzzy logic and fractal theory for applications in medical diagnosis. The main idea is that a higher type and order of fuzzy logic can help in solving various diagnosis problems and finding better results. In addition, fractal theory is also employed for enhancing medical diagnosis. In this regard, several hybrid intelligent methods are offered. In this book, we test the proposed methods using several medical diagnosis problems, like diagnosis of problems in the brain, heart, lungs, and others. We can notice that when type-3 fuzzy systems are implemented to model the behavior of systems, the results in diagnosis are enhanced, because the management of uncertainty is better. For this reason, we consider in this book the proposed methods using type-3 fuzzy systems and fractal theory to improve the diagnosis in complex medical problems.

This book is intended to be a reference for scientists and engineers interested in applying type-3 fuzzy logic and fractal theory techniques in medical diagnosis. This book can also be used as a reference for graduate courses like the following: soft computing, fuzzy logic, neural networks, bio-inspired algorithms, intelligent prediction, and similar ones. We consider that this book can also be used to get novel ideas for new lines of research or to continue the lines of research proposed by the authors of the book.

In Chap. 1, we start by offering a brief introduction of the potential use of type-3 fuzzy systems and fractal theory in complex medical diagnosis problems. We also mention other possible applications of the proposed prediction approach.

We describe in Chap. 2 a type-3 fuzzy-fractal approach for tuberculosis diagnosis. In this approach, the fractal dimension captures the geometrical complexity of the lungs and a type-3 fuzzy system contains the knowledge of medical expert in the diagnosis.

Chapter 3 is devoted to describe a combination of type-3 fuzzy logic system and the fractal dimension is employed for osteoporosis diagnosis. The fractal dimension is utilized to characterize the geometrical properties of the bone texture and anisotropy.

We describe in Chap. 4 a hybrid combination of type-3 fuzzy logic and the fractal dimension is employed for Leukemia cancer diagnosis. The fractal dimension is utilized to characterize the geometrical properties in the blood samples of a patient. A type-3 fuzzy system is employed to represent the knowledge of experts in diagnosis.

We offer in Chap. 5 a combination of type-3 fuzzy logic system and the fractal dimension is employed for diagnosis of vascular diseases. Fractal dimension is utilized to characterize the geometrical properties of the cardiac vessels. The type-3 fuzzy system is employed to represent the knowledge of experts in vascular diagnosis.

We describe in Chap. 6 a hybrid combination of type-3 fuzzy logic and the Holder exponent, which measures multi-fractality, is utilized for diagnosis of mental disorders. The Holder exponent is utilized to characterize the geometrical properties of the EEG signal. A type-3 fuzzy system is employed to represent the knowledge of experts in diagnosis of mental disorders.

We explain in Chap. 7 an approach with type-3 fuzzy logic system and the fractal dimension is employed for diagnosis of vascular cerebral diseases. Fractal theory is utilized to characterize the geometrical properties of the cerebral vessels. A type-3 fuzzy system is employed to represent the knowledge of experts in vascular cerebral diagnosis.

We describe in Chap. 8 the conclusions of this work on type-3 fuzzy logic and fractal theory for medical diagnosis, as well as some future research works are envisioned.

We end this preface of the book by giving thanks to all the people who have helped or encouraged us during the writing of this book. First of all, we would like to thank our colleague and friend Prof. Juan Ramón Castro for always supporting our work, and for motivating us to write our research work. We would also like to thank our colleagues working in soft computing, which are too many to mention each by their name. Of course, we need to thank our supporting agencies, CONACYT and TecNM, in our country for their help during this project. We have to thank our institution, Tijuana Institute of Technology, for always supporting our projects. Finally, we thank our respective families for their continuous support during the time that we spent in this project.

Baja California, Mexico
July 2024

Prof. Oscar Castillo
Prof. Patricia Melin

Contents

1 Introduction to Type-3 Fuzzy-Fractal Medical Diagnosis 1
   References 2
2 A Type-3 Fuzzy-Fractal Approach for Tuberculosis Diagnosis 5
   2.1 Introduction 5
   2.2 Type-3 Fuzzy Theory 6
   2.3 Fractal Dimension 6
   2.4 Type-1 Fuzzy-Fractal Approach 6
   2.5 Proposed Type-3 Approach 8
   2.6 Application to Tuberculosis Diagnosis 11
   2.7 Conclusions 12
   References 12
3 A Type-3 Fuzzy-Fractal Approach for Bone Analysis
   in Osteoporosis Diagnosis 15
   3.1 Introduction 15
   3.2 Type-3 Fuzzy Theory 16
   3.3 Fractal Dimension 16
   3.4 Type-1 Fuzzy-Fractal Approach 17
   3.5 Proposed Type-3 Approach 19
   3.6 Application to Osteoporosis Diagnosis 20
   3.7 Conclusions 21
   References 24
4 A Type-3 Fuzzy-Fractal Approach for Leukemia Cancer
   Diagnosis 27
   4.1 Introduction 27
   4.2 Type-3 Fuzzy Logic 28
   4.3 Fractal Dimension 28
   4.4 Type-1 Fuzzy-Fractal Approach 28
   4.5 Proposed Type-3 Approach 30
   4.6 Application to Leukemia Diagnosis 33

	4.7 Conclusions	33
	References	34
5	**A Type-3 Fuzzy-Fractal Approach for Diagnosis of Vascular Diseases Based on Cardiac Vessels**	**37**
	5.1 Introduction	37
	5.2 Type-3 Fuzzy Theory	38
	5.3 Fractal Dimension	38
	5.4 Type-1 Fuzzy-Fractal Approach	39
	5.5 Proposed Type-3 Approach	40
	5.6 Application to Brain Tumor Diagnosis	43
	5.7 Conclusions	43
	References	44
6	**A Type-3 Fuzzy-Fractal Approach for Diagnosis of Mental Disorders**	**47**
	6.1 Introduction	47
	6.2 Type-3 Fuzzy Logic	48
	6.3 Fractal Dimension and Holder Exponents	48
	6.4 Type-1 Fuzzy-Fractal Approach	49
	6.5 Proposed Type-3 Approach	50
	6.6 Application to Mental Disorder Diagnosis	53
	6.7 Conclusions	54
	References	54
7	**A Type-3 Fuzzy-Fractal Approach for Diagnosis of Vascular Diseases Based on Cerebral Vessels**	**57**
	7.1 Introduction	57
	7.2 Type-3 Fuzzy Theory	58
	7.3 Fractal Dimension	58
	7.4 Type-1 Fuzzy-Fractal Approach	59
	7.5 Proposed Type-3 Approach	60
	7.6 Application to Diagnosis of Cerebral Vascular Diseases	63
	7.7 Conclusions	63
	References	64
8	**Conclusions of Type-3 Fuzzy-Fractal Medical Diagnosis**	**67**
	References	68
Index		**71**

Chapter 1
Introduction to Type-3 Fuzzy-Fractal Medical Diagnosis

Type-3 fuzzy logic (T3FL) and fractal theory (FT) and their use in medical diagnosis have been put forward in this book. The focus is on the essential reasons for employing type-3 in achieving an accurate medical diagnosis. Type-3 is a new theory to model uncertainty that can be employed in diagnosis. Type-2 has been previously employed as a way for dealing with medical diagnosis, but recently type-3 offers an alternative in dealing with more complex medical problems. In this work, we briefly review the theory of type-3, which could be studied in a more profound way. Additionally, we postulate a way for building type-3 fuzzy-fractal systems in medical diagnosis.

The concept of information is inherently associated with the concept of uncertainty [1, 2]. Uncertainty is an attribute of information [3]. For systems based on type-1 sets, an uncertainty model with values in [0, 1] is employed. When an entity is uncertain, like a measurement, it is difficult to specify its exact value, and of course a type-1 fuzzy set makes more sense than a traditional set [3, 4]. However, it is not reasonable to use an accurate membership function for something uncertain, so in this case what we need is another type of fuzzy sets, those which are able to handle these uncertainties, and the so-called type-2 fuzzy sets [5, 6] were put forward for achieving this purpose. The amount of uncertainty in a system can be handled by using type-2 fuzzy logic because this logic offers better capabilities to handle linguistic uncertainties by modeling vagueness and unreliability of information [7, 8]. Recently, there has been some works dealing with interval type-3 fuzzy models, as they can provide even better capabilities for handling uncertainty in medical applications [9] and other areas, and this is the main motivation of this book.

Type-2 fuzzy models have emerged as an interesting generalization of fuzzy models based upon type-1 fuzzy sets [5, 10]. There have been a number of claims put forward as to the relevance of type-2 fuzzy sets being regarded as generic building constructs of fuzzy models [11–13]. Likewise, there is a record of some experimental evidence showing some improvements in terms of accuracy of fuzzy models of type-2 over their type-1 counterparts [14–18]. There have been a lot of applications of type-2 in intelligent control [19–24], pattern recognition [25–29], intelligent manufacturing [8, 16, 30], time series prediction [14, 31], and others [8]. Recently, there has been some evidence [9] that type-3 fuzzy systems can improve results with respect to type-2 in some cases, and for this reason the importance of this book that provides the basic concepts to develop these type-3 fuzzy-fractal systems for medical diagnosis.

We are considering several medical diagnosis applications in this book, but we plan to consider in the future more complex diagnosis problems, such as in [32–35], and other kinds of problems, such as the ones presented in [36, 37].

The rest of the book is formed by seven chapters, which are mentioned below. Chapter 2 deals with tuberculosis diagnosis based on the fractal dimension and T3FL. Chapter 3 outlines the type-3 fuzzy-fractal approach for bone analysis and osteoporosis diagnosis. Chapter 4 outlines the approach for Leukemia cancer diagnosis based on type-3 and FT. Chapter 5 describes the diagnosis for vascular diseases in cardiac vessels. Chapter 6 delineates an approach for diagnosis of mental disorders combining type-3 and FT. Chapter 7 describes the diagnosis for vascular diseases in cerebral vessels. Finally, Chapt. 8 outlines the conclusions. Lastly, we can state that this book will be a reference on the type-3 fuzzy-fractal approach for those interested in utilizing it, but can also could serve as an inspiration to researchers in contributing to this relevant area of medical diagnosis.

References

1. P. Melin, O. Castillo, *Modelling, Simulation and Control of Non-Linear Dynamical Systems* (Taylor and Francis, London, Great Britain, 2002)
2. J.M. Mendel, Uncertainty, fuzzy logic, and signal processing. Signal Process. J. **80**, 913–933 (2000)
3. L.A. Zadeh, The concept of a linguistic variable and its application to approximate reasoning. Inf. Sci. **8**, 43–80 (1975)
4. J.R. Jang, C.T. Sun, E. Mizutani, *Neuro-Fuzzy and Soft Computing* (Prentice Hall, Upper Saddle River, NJ, USA, 1997)
5. O. Castillo, P. Melin, *Type-2 Fuzzy Logic: Theory and Applications* (Springer-Verlag, Heidelberg, Germany, 2008)
6. N. N. Karnik, J. M. Mendel, An Introduction to Type-2 Fuzzy Logic Systems. Technical Report. University of Southern California (1998)
7. M. Wagenknecht, K. Hartmann, Application of fuzzy sets of type 2 to the solution of fuzzy equations systems. Fuzzy Sets Syst. **25**, 183–190 (1988)
8. M.H.F. Zarandi, I.B. Turksen, O.T. Kasbi, Type-2 fuzzy modelling for desulphurization of steel process. Expert Syst. Appl. **32**, 157–171 (2007)
9. A. Mohammadzadeh, O. Castillo, S.S. Band et al., A novel fractional-order multiple-model type-3 fuzzy control for nonlinear systems with unmodeled dynamics. Int. J. Fuzzy Syst. **23**, 1633–1651 (2021)
10. H. Hagras, Hierarchical type-2 fuzzy logic control architecture for autonomous mobile robots. IEEE Trans. Fuzzy Syst. **12**, 524–539 (2004)
11. S. Coupland, R. John, New geometric inference techniques for type-2 fuzzy sets. Int. J. Approximate Reasoning **49**, 198–211 (2008)
12. J.T. Starczewski, Efficient triangular type-2 fuzzy logic systems. Int. J. Approx. Reason. **50**, 799–811 (2009)

13. C. Walker, E. Walker, Sets with type-2 operations. Int. J. Approx. Reason. **50**, 63–71 (2009)
14. N. S. Bajestani, A. Zare, Application of optimized type-2 fuzzy time series to forecast Taiwan stock index, in *2nd International Conference on computer, Control and Communication* (2009), pp. 275–280
15. J.R. Castro, O. Castillo, P. Melin, A. Rodriguez-Diaz, A hybrid learning algorithm for a class of interval type-2 fuzzy neural networks. Inf. Sci. **179**, 2175–2193 (2009)
16. T. Dereli, A. Baykasoglu, K. Altun, A. Durmusoglu, I.B. Turksen, Industrial applications of type-2 fuzzy sets and systems: a concise review. Comput. Ind. **62**, 125–137 (2011)
17. C. Leal-Ramirez, O. Castillo, P. Melin, A. Rodriguez-Diaz, Simulation of the bird age-structured population growth based on an interval type-2 fuzzy cellular structure. Inf. Sci. **181**, 519–535 (2011)
18. R. Martinez, O. Castillo, L. T. Aguilar, Optimization of interval type-2 fuzzy logic controllers for a perturbed autonomous wheeled mobile robot using genetic algorithms. Inf. Sci. **179**(13), 2158–2174 (2009)
19. M. Hsiao, T.H.S. Li, J.Z. Lee, C.H. Chao, S.H. Tsai, Design of interval type-2 fuzzy sliding-mode controller. Inf. Sci. **178**(6), 1686–1716 (2008)
20. P. Melin, O. Castillo, A new method for adaptive model-based control of non-linear dynamic plants using a neuro-fuzzy-fractal approach, Journal of. Soft. Comput. **5**, 171–177 (2001)
21. P. Melin, O. Castillo, A new method for adaptive model-based control of nonlinear plants using type-2 fuzzy logic and neural networks, in *Proceedings IEEE FUZZ Conference* (2003), pp. 420–425
22. T. Ozen, J.M. Garibaldi, Investigating adaptation in type-2 fuzzy logic systems applied to umbilical acid-base assessment, in *European Symposium on Intelligent Technologies, Hybrid Systems and their implementation on Smart Adaptive Systems (EUNITE 2003)* (Oulu, Finland, 2003)
23. R. Sepulveda, O. Castillo, P. Melin, O. Montiel, An efficient computational method to implement type-2 fuzzy logic in control applications. Adv. Soft Comput. **41**, 45–52 (2007)
24. R. Sepulveda, O. Castillo, P. Melin, A. Rodriguez-Diaz, O. Montiel, Experimental study of intelligent controllers under uncertainty using type-1 and type-2 fuzzy logic. Inf. Sci. **177**(10), 2023–2048 (2007)
25. P. Melin, O. Castillo, *Hybrid Intelligent Systems for Pattern Recognition* (Springer-Verlag, Heidelberg, Germany, 2005)
26. O. Mendoza, P. Melin, O. Castillo, G. Licea, *Type-2 Fuzzy Logic for Improving Training Data and Response Integration in Modular Neural Networks for Image Recognition*. Lecture Notes in Artificial Intelligence, vol. 4529 (2007), pp. 604–612
27. O. Mendoza, P. Melin, O. Castillo, Interval type-2 fuzzy logic and modular neural networks for face recognition applications. Appl. Soft Comput. J. **9**, 1377–1387 (2009)
28. O. Mendoza, P. Melin, G. Licea, Interval type-2 fuzzy logic for edges detection in digital images. Int. J. Intell. Syst. **24**, 1115–1133 (2009)
29. J. Urias, D. Hidalgo, P. Melin, O. Castillo, A method for response integration in modular neural networks with type-2 fuzzy logic for biometric systems. Adv. Soft Comput. **41**, 5–15 (2007)
30. P. Melin, O. Castillo, An intelligent hybrid approach for industrial quality control combining neural networks, fuzzy logic and fractal theory. Inf. Sci. **177**, 1543–1557 (2007)
31. O. Castillo, P. Melin, Hybrid intelligent systems for time series prediction using neural networks, fuzzy logic and fractal theory. IEEE Trans. Neural Netw. **13**, 1395–1408 (2002)
32. O. Castillo, J.R. Castro, P. Melin, Interval type-3 fuzzy aggregation of neural networks for multiple time series prediction: the case of financial forecasting. Axioms **11**, 251 (2022). https://doi.org/10.3390/axioms11060251
33. M. Ramirez, P. Melin, A new perspective for multivariate time series decision making through a nested computational approach using type-2 fuzzy integration. Axioms **12**, 385 (2023). https://doi.org/10.3390/axioms12040385
34. M. Ramírez, P. Melin, O. Castillo, Interval type-3 fuzzy aggregation for hybrid-hierarchical neural classification and prediction models in decision-making. Axioms **12**, 906 (2023). https://doi.org/10.3390/axioms12100906

35. P. Melin, D. Sánchez, J.R. Castro, O. Castillo, Design of type-3 fuzzy systems and ensemble neural networks for covid-19 time series prediction using a firefly algorithm. Axioms **11**, 410 (2022). https://doi.org/10.3390/axioms11080410
36. E. Ontiveros, P. Melin, O. Castillo, Comparative study of interval type-2 and general type-2 fuzzy systems in medical diagnosis. Inf. Sci. **525**, 37–53 (2020)
37. F. Valdez, J.C. Vazquez, P. Melin, O. Castillo, Comparative study of the use of fuzzy logic in improving particle swarm optimization variants for mathematical functions using co-evolution. Appl. Soft Comput. **52**, 1070–1083 (2017)

Chapter 2
A Type-3 Fuzzy-Fractal Approach for Tuberculosis Diagnosis

In this chapter a combination of type-3 fuzzy logic system (T3FLS) and the fractal dimension (FD) is employed for tuberculosis diagnosis. FD is utilized to characterize the geometrical properties of a region in lung X-ray images. A T3FLS is employed to represent the knowledge of experts in tuberculosis diagnosis. In this case, type-3 fuzzy logic helps in representing the uncertainty in making a diagnosis. Simulation results with a dataset of lung X-rays shows the accuracy of the proposal.

2.1 Introduction

Now it is accepted that fuzzy systems achieve remarkable applications in many areas, like: forecasting and plant monitoring. Originally, fuzzy sets were offered by Zadeh in 1965 [1, 2]. Later fuzzy logic and systems were also offered by Zadeh, and many applications follow, like in [3, 4]. Fuzzy Logic has evolved from the initial studies of Zadeh with type-1 fuzzy theory [1], to later the theory of type-2 [5–7], to now where type-3 is appearing [8–10]. Now even type-n has been mentioned [11]. In addition, type-3 has been employed to handle uncertainty from measurements, like in control [12–17]. There have been some previous works applying the fractal dimension (FD) to diagnosis in diverse medical problems, as in [18, 19]. Also, there exist some works using fuzzy logic in medical diagnosis, like in [20, 21]. Unlike previous works, in this paper we are offering for the first time a type-3 fuzzy-fractal hybrid for tuberculosis diagnosis. We employ x-ray images of a dataset obtained in [22] for testing our approach with images for normal and tuberculosis people [22].

The rest of the chapter is: Sect. 2.2 is summarizing type-3 terminology, in Sect. 2.3 the fractal dimension concept is offered, in Sect. 2.4 the type-1 fuzzy fractal diagnosis is delineated, in Sect. 2.5 we are offering the type-3 fuzzy fractal diagnosis approach, in Sect. 2.6 results are presented, and in Sect. 2.7 conclusions are offered.

2.2 Type-3 Fuzzy Theory

We first postulate the concepts.

Definition 2.1 A type-3 fuzzy set (T3 FS) [8], written as $A^{(3)}$, is represented by the membership function (MF) of $A^{(3)}$, in the Cartesian product $X \times [0, 1] \times [0, 1]$ in $[0, 1]$, where X is the primary variable universe of $A^{(3)}$, x. The MF of $\mu_{A^{(3)}}$ is a type-3 MF (T3 MF):

$$\mu_{A^{(3)}} : X \times [0, 1] \times [0, 1] \to [0, 1]$$
$$A^{(3)} = \{(x, u(x), v(x, u), \mu_{A^{(3)}}(x, u, v)) | x \in X, u \in U \subseteq [0, 1], v \in V \subseteq [0, 1]\} \quad (2.1)$$

where U is universe for secondary variable u and V is universe for tertiary variable v. When T3MFs are employed in the fuzzy rules then we have a type-3 fuzzy logic system (T3FLS).

2.3 Fractal Dimension

Recently, remarkable advances have been achieved in understanding the complexity of an object by employing fractal theory [18, 19]. For example, time series in finance suggest fractal structure. The FD is postulated as:

$$d = \lim_{r \to 0} [lnN(r)]/[ln(1/r)] \quad (2.2)$$

where $N(r)$ is for number of boxes of size r. The d defined in (2.2) is approximated with box covering for several r sizes and then employing regression for finding the d value.

2.4 Type-1 Fuzzy-Fractal Approach

We are presenting a fuzzy fractal approach for tuberculosis diagnosis. In Fig. 2.1 we can notice the fuzzy system structure, in which the fractal dimension (FD) of the x-ray image enters as input, and the diagnosis is the output.

In Fig. 2.2 we can find the MFs for the FD, which are Low and High. In Fig. 2.3 we can find the MFs for the output, which is the diagnosis (DIAG). In this case, there are two MFs, one for Tuberculosis (T) and the other for Normal (N).

2.4 Type-1 Fuzzy-Fractal Approach

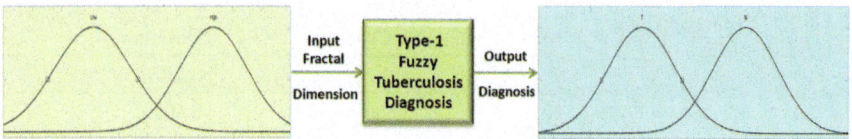

Fig. 2.1 Structure of the fuzzy fractal approach for tuberculosis diagnosis

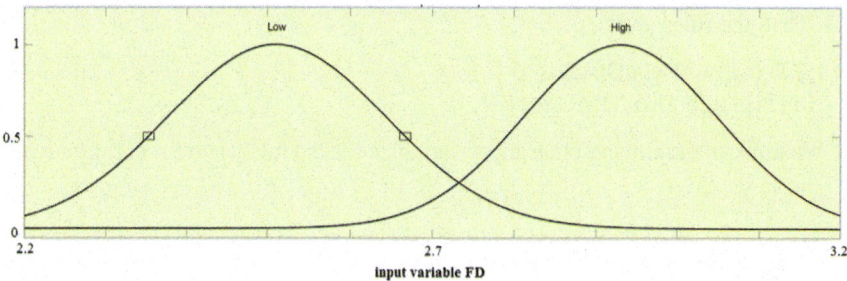

Fig. 2.2 MFs for the FD input

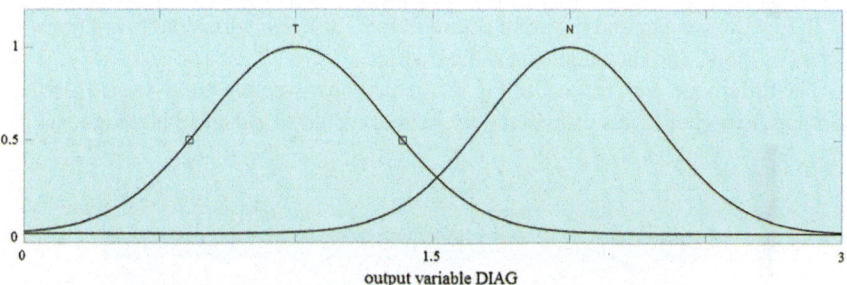

Fig. 2.3 MFs for the DIAG output variable

Table 2.1 summarizes the MF parameters for the FD input. The range for the FD is from 2.2 to 3.2, as experimentally it has been found that the FDs of the x-rays are in this range. Table 2.2 presents the parameters for the MFs of DIAG, which we consider to be in the range from 0 to 3. All MFs are Gaussian, so we have 2 parameters, center and deviation.

Table 2.1 Parameter values of the input MFs

MFs	Center	Deviation
Low	2.51	0.1139
High	2.93	0.1541

Table 2.2 Parameter values of the output MFs

MFs	Center	Deviation
T	1.0	0.2270
N	2.0	0.4058

Based on expert knowledge and experimentation with the x-ray dataset it is known that tuberculosis corresponds to low values of FD, and normal is for high values of FD. Then the rules are:

If FD is Low Then DIAG is T
If FD is High Then DIAG is N

We think of employing other input variables for the fuzzy system in future work.

2.5 Proposed Type-3 Approach

We now elevate the type-1 fuzzy system to type-3 by employing T3MFs. Now the input T3MFs are depicted in Fig. 2.4. The output T3MFs are illustrated in Fig. 2.5.

In Fig. 2.6 we can find the architecture of the T3FLS in which the FD of the x-ray enters as input, and the diagnosis is the output.

The difference with respect to Fig. 2.1 is that now we are employing T3MFs that can handle higher levels of uncertainty, which would in enhance the diagnosis. We

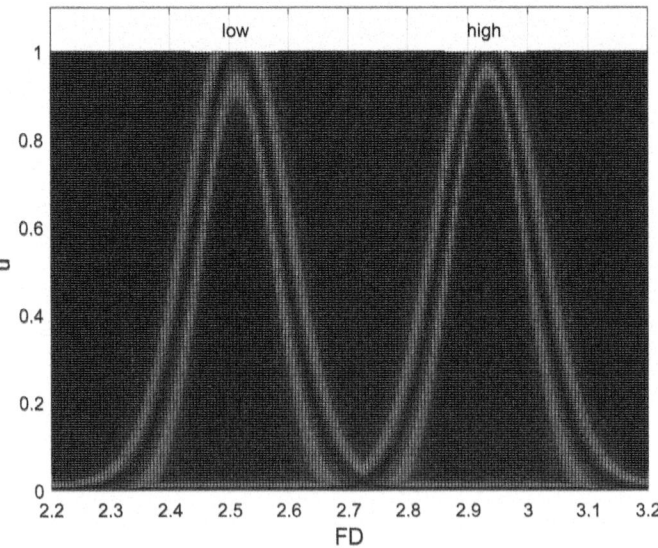

Fig. 2.4 Input T3MFs

2.5 Proposed Type-3 Approach

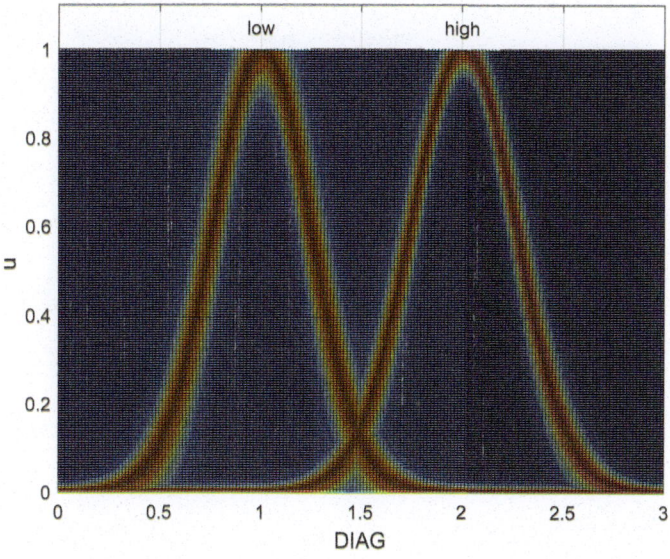

Fig. 2.5 Output T3MFs

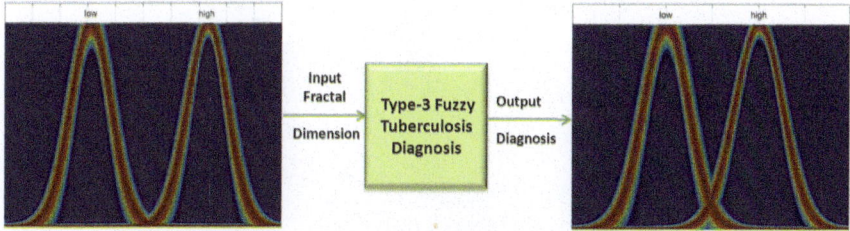

Fig. 2.6 Architecture of the T3FLS for tumor diagnosis

also show in Figs. 2.7 and 2.8 a three-dimensional (3-D) view of T3MFs for the input and output, respectively.

Table 2.3 summarizes the parameter values for the T3MFs of the FD input. Table 2.4 lists the parameters for the T3MFs of DIAG, which we used in the range from 0 to 3. The T3MFs are scaled Gaussians, so we now have four parameters: center, deviation, lower lag and lower scale [8].

The rules are the same, as now the difference is that the T3FLS employs T3MFs.

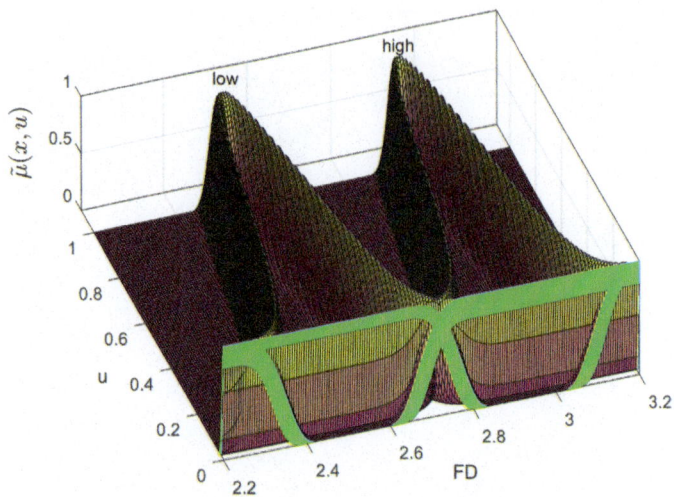

Fig. 2.7 3-D view of Input T3MFs

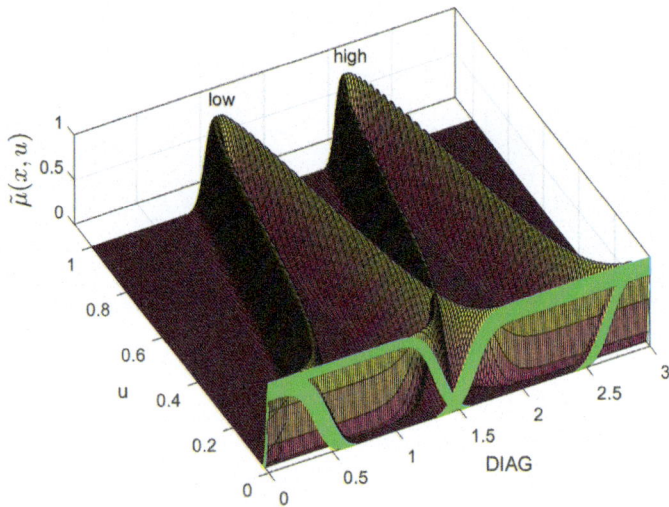

Fig. 2.8 3-D view of output T3MFs

Table 2.3 Parameter values of the input T3MFs

MFs	Center	Deviation	Lower scale	Lower Lag
Low	2.51	0.09	0.8	0.2
High	2.93	0.09	0.9	0.2

Table 2.4 Parameter values of the output T3MFs

MFs	Center	Deviation	Lower scale	Lower Lag
Low	1.0	0.3	0.8	0.2
High	2.0	0.3	0.9	0.1

2.6 Application to Tuberculosis Diagnosis

This section is presenting the results of the fuzzy-fractal systems for tuberculosis diagnosis. Table 2.5 summarizes a comparison of results, where column one shows the FD values for 20 cases, column two lists the corresponding diagnosis of experts (T = 1, N = 2), column three shows the results of type-1, and column four results of type-3. From Table 2.5 we note that the type-3 fuzzy-fractal approach produces 100% correct diagnosis in 20 cases, however numerically the results of type-1 only 90% correct values, with respect to experts.

Table 2.5 Comparison of fuzzy-fractal approaches for tuberculosis diagnosis

FD	DIAG	Type-1 Fuzzy DIAG	Fuzzy Type-3 DIAG
2.990	N	1.99 (N)	1.9894 (N)
2.403	T	1.01 (T)	1.0177 (T)
2.964	N	1.99 (N)	1.9905 (N)
2.552	T	1.14 (T)	1.0096 (T)
2.510	T	1.07 (T)	1.0091 (T)
2.515	T	1.08 (T)	1.0091 (T)
2.870	N	1.99 (N)	1.9894 (N)
2.673	T	1.60 (N)	1.0720 (T)
2.917	N	1.99 (N)	1.9907 (N)
2.688	T	1.67 (N)	1.1291 (T)
2.997	N	1.99 (N)	1.9888 (N)
2.592	T	1.25 (T)	1.0128 (T)
2.907	N	1.99 (N)	1.9906 (N)
2.992	N	1.99 (N)	1.9892 (N)
2.416	T	1.01 (T)	1.0147 (T)
2.353	T	1.00 (T)	1.0498 (T)
2.976	N	1.99 (N)	1.9901 (N)
2.427	T	1.02 (T)	1.0129 (T)
2.880	N	1.99 (N)	1.9899 (N)
2.404	T	1.01 (T)	1.174 (T)

2.7 Conclusions

In this chapter a hybrid of type-3 and FD was employed for tuberculosis diagnosis. FD was utilized to characterize the geometrical properties of the x-ray image. A type-3 system was employed to convey the knowledge of experts in diagnosis. Simulation results with a dataset of x-rays shows the effectiveness of the proposal. In future work we could employ the diagnosis approach for other medical problems. In addition, we could consider optimizing the parameters of the fuzzy systems by utilizing metaheuristics, such as the ones presented in [23–26].

References

1. L.A. Zadeh, Fuzzy sets. Inf. Control **8**, 338–353 (1965)
2. L.A. Zadeh, Knowledge representation in fuzzy logic. IEEE Trans. Knowl. Data Eng. **1**, 89 (1989)
3. P. Melin, O. Castillo, Adaptive intelligent control of aircraft systems with a hybrid approach combining neural networks, fuzzy logic and fractal theory. Appl. Soft Comput. **3**(4), 353–362 (2003)
4. O. Castillo, P. Melin, Intelligent adaptive model-based control of robotic dynamic systems with a hybrid fuzzy-neural approach. Appl. Soft Comput. **3**(4), 363–378 (2003)
5. J.M. Mendel, R.I. Bob John, Type-2 fuzzy sets made simple. IEEE Trans. Fuzzy Syst. **10**(2), 117–127 (2002)
6. L Astudillo, O Castillo, P Melin, A Alanis, J Soria, LT Aguilar, Intelligent control of an autonomous mobile robot using type-2 fuzzy logic. Eng. Lett. **13**(3) (2006)
7. R. Sepúlveda, O. Montiel, O. Castillo, P. Melin, Embedding a high speed interval type-2 fuzzy controller for a real plant into an FPGA. Appl. Soft Comput. **12**(3), 988–998 (2012)
8. O. Castillo, J.R. Castro, P. Melin, *Interval Type-3 Fuzzy Systems: Theory and Design* (Springer, Cham, Switzerland, 2022)
9. O. Castillo, P. Melin, Towards interval type-3 intuitionistic fuzzy sets and systems. Mathematics **10**, 4091 (2022)
10. M.W. Tian, S.R. Yan, A. Mohammadzadeh, J. Tavoosi, S. Mobayen, R. Safdar, W. Assawinchaichote, M.T. Vu, A. Zhilenkov, Stability of interval type-3 fuzzy controllers for autonomous vehicles. Mathematics **9**(21), 2742 (2021)
11. J.T. Rickard, J. Aisbett, G. Gibbon, Fuzzy subsethood for fuzzy sets of type-2 and generalized type-n. IEEE Trans. Fuzzy Syst. **17**(1), 50–60 (2009)
12. A. Mohammadzadeh, M.H. Sabzalian, W. Zhang, An interval type-3 fuzzy system and a new online fractional-order learning algorithm: theory and practice. IEEE Trans. Fuzzy Syst. **28**(9), 1940–1950 (2020)
13. Z. Liu, A. Mohammadzadeh, H. Turabieh, M. Mafarja, S.S. Band, A. Mosavi, A new online learned interval type-3 fuzzy control system for solar energy management systems. IEEE Access **9**, 10498–10508 (2021)
14. D. Singh, N.K. Verma, A.K. Ghosh, A.K. Malagaudanavar, An approach towards the design of interval type-3 TS fuzzy system. IEEE Trans. Fuzzy Syst. (2021)
15. J.H. Wang, J. Tavoosi, A. Mohammadzadeh, S. Mobayen, J.H. Asad, W. Assawinchaichote, P. Skruch, Non-singleton type-3 fuzzy approach for flowmeter fault detection: experimental study in a gas industry. Sensors **21**(21), 7419 (2021)
16. K.A. Alattas, A. Mohammadzadeh, S. Mobayen, A.A. Aly, B.F. Felemban, A new data-driven control system for MEMSs gyroscopes: dynamics estimation by type-3 fuzzy systems. Micromachines **12**(11), 1390 (2021)

References

17. Y. Cao, A. Raise, A. Mohammadzadeh, S. Rathinasamy, S.S. Band, A. Mosavi, Deep learned recurrent type-3 fuzzy system: application for renewable energy modeling/prediction. Energy Rep. **7**, 8115–8127 (2021)
18. A. Chan, J.A. Tuszynski, Automatic prediction of tumour malignancy in breast cancer with fractal dimension. R. Soc. Open Sci. **3**(12), 160558 (2016)
19. Y.D. Zhang, X.Q. Chen, T.M. Zhan, Z.Q. Jiao, Y. Sun, Z.M. Chen, S.H. Wang, Fractal dimension estimation for developing pathological brain detection system based on Minkowski-Bouligand method. IEEE Access **4**, 5937–5947 (2016)
20. O. Alpar, A mathematical fuzzy fusion framework for whole tumor segmentation in multimodal MRI using Nakagami imaging. Expert Syst. Appl. **216**, 119462 (2023)
21. O. Alpar, R. Dolezal, P. Ryska, O. Krejcar, Nakagami-Fuzzy imaging framework for precise lesion segmentation in MRI. Pattern Recogn. **128**, 108675 (2022)
22. G. Abdikerimova, A. Shekerbek, M. Tulenbayev, B. Sultanova, S. Beglerova, E. Dzhaulybaeva, K. Zhumakanova, B. Rysbekkyzy, Detection of lung pathology using the fractal method. Int. J. Electr. Comput. Eng. **13**(6), 6778–6786 (2023)
23. B. González, F. Valdez, P. Melin, G. Prado-Arechiga, Fuzzy logic in the gravitational search algorithm for the optimization of modular neural networks in pattern recognition. Expert Syst. Appl. **42**(14), 5839–5847 (2015)
24. L. Amador-Angulo, O. Mendoza, J.R. Castro, A. Rodriguez-Diaz, P. Melin, O. Castillo, Fuzzy sets in dynamic adaptation of parameters of a bee colony optimization for controlling the trajectory of an autonomous mobile robot. Sensors **16**(9), 1458 (2016)
25. O. Castillo, E. Lizarraga, J. Soria, P. Melin, F. Valdez, New approach using ant colony optimization with ant set partition for fuzzy control design applied to the ball and beam system. Inf. Sci. **294**, 203–215 (2015)
26. R. Martínez-Soto, O. Castillo, L.T. Aguilar, Type-1 and Type-2 fuzzy logic controller design using a hybrid PSO–GA optimization method. Inf. Sci. **285**, 35–49 (2014)

Chapter 3
A Type-3 Fuzzy-Fractal Approach for Bone Analysis in Osteoporosis Diagnosis

In this chapter a combination of type-3 fuzzy logic system (T3FLS) and the fractal dimension (FD) is employed for osteoporosis diagnosis. FD is utilized to characterize the geometrical properties of the bone texture and anisotropy. A T3FLS is employed to represent the knowledge of experts in diagnosis. In this situation, the T3FLS helps in representing the uncertainty in making a medical diagnosis. Simulation results with a dataset of bone x-rays shows the accuracy of the proposal.

3.1 Introduction

Fuzzy systems have produced remarkable applications in many areas, like: control, forecasting and plant monitoring. Originally, fuzzy sets offered by Zadeh in 1965 [1, 2]. Later fuzzy logic and systems were also offered by Zadeh, and many applications follow, like in [3, 4]. Fuzzy Logic has evolved from the initial studies of Zadeh with type-1 fuzzy theory [1], to later the theory of type-2 [5–7], to now where type-3 is appearing [8–10]. Now even type-n has been mentioned [11]. In addition, type-3 has been employed to handle uncertainty from measurements, like in control [12–17]. There have been some previous works applying the fractal dimension (FD) to diagnosis in diverse medical problems, as in [18, 19]. Also, there exist some works using fuzzy logic in medical diagnosis, like in [20–23]. Unlike previous works, in this paper we are offering for the first time a type-3 fuzzy-fractal hybrid for osteoporosis diagnosis. We employ information from a dataset of healthy and osteoporosis cases, for which the fractal analysis was previously made [24], for testing our approach. Some examples of images from this dataset, showing Osteoporosis cases in different locations of the human body are shown in Fig. 3.1.

The rest of the chapter is: Sect. 3.2 is briefly summarizing type-3 terminology, in Sect. 3.3 the fractal dimension concept is offered, in Sect. 3.4 the type-1 fuzzy fractal diagnosis is delineated, in Sect. 3.5 we are offering the type-3 fuzzy fractal

Fig. 3.1 Sample image of Osteoporosis in the hand

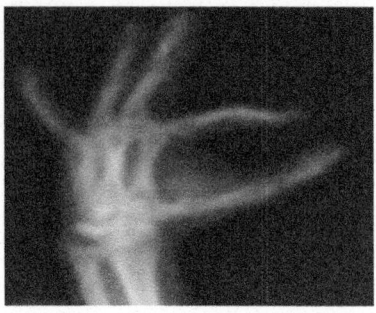

diagnosis approach, in Sect. 3.6 results are listed, and in Sect. 3.7 conclusions are offered.

3.2 Type-3 Fuzzy Theory

We first postulate the concepts.

Definition 3.1 A type-3 fuzzy set (T3 FS) [8], written as $A^{(3)}$, is represented by the membership function (MF) of $A^{(3)}$, in the Cartesian product $X \times [0, 1] \times [0, 1]$ in $[0, 1]$, where X is the primary variable universe of $A^{(3)}$, x. The MF of $\mu_{A^{(3)}}$ is a type-3 MF (T3 MF):

$$\mu_{A^{(3)}} : X \times [0, 1] \times [0, 1] \to [0, 1]$$
$$A^{(3)} = \{(x, u(x), v(x, u), \mu_{A^{(3)}}(x, u, v)) | x \in X, u \in U \subseteq [0, 1], v \in V \subseteq [0, 1]\} \tag{3.1}$$

where U is universe for secondary variable u and V is universe for tertiary variable v. When T3MFs are employed in the fuzzy rules then we have a type-3 fuzzy logic system (T3FLS).

3.3 Fractal Dimension

Recently, remarkable advances have been achieved in understanding the complexity of an object by employing fractal theory [18, 19]. For example, time series of the brain suggest fractal structure. The FD is postulated as:

$$d = \lim_{r \to 0}[lnN(r)]/[ln(1/r)] \tag{3.2}$$

where $N(r)$ is for number of boxes of size r. The d defined in (3.2) is approximated with box covering for several r sizes and then employing regression for finding the d value.

3.4 Type-1 Fuzzy-Fractal Approach

We are presenting a fuzzy fractal approach for osteoporosis diagnosis. Bone strength depends on bone mass and the internal trabecular bone structure. A fractal analysis characterizes the roughness (C0) and anisotropy (ANI) of the bone and can determine if there can be osteoporosis. In Fig. 3.2 we can notice the fuzzy system structure, in which the fractal dimension (FD) of the roughness (FD_{C0}) and anisotropy (F_{ANI}) enter as inputs, and the diagnosis is the output.

In Fig. 3.3 we can find the MFs for the FD_{C0}, which are Low and High. In Fig. 3.4 we show the MFs for the, which are also Low and High. In Fig. 3.5 we can find the MFs for the output, which is the diagnosis (DIAG). In this case, there are two MFs, one for Healthy (He) and the other for Osteoporosis (Os).

Table 3.1 summarizes the MF parameters for the FD_{C0} input. The range for the FD is from 0.7 to 1.5, as experimentally it has been found that the FDs of the images of healthy and osteoporosis bones are in this range [25]. Table 3.2 summarizes the MF parameters for the FD_{ANI} input. The range for the FD is from 0.7 to 1.0, as experimentally it has been found that the FDs of the images for this parameter are in this range. Table 3.3 presents the parameters for the MFs of DIAG, which we consider

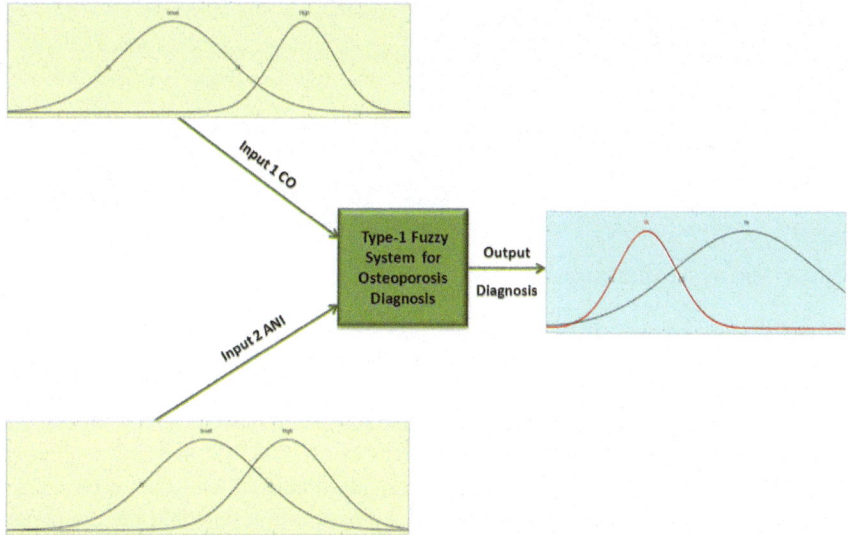

Fig. 3.2 Structure of the approach for Osteoporosis diagnosis

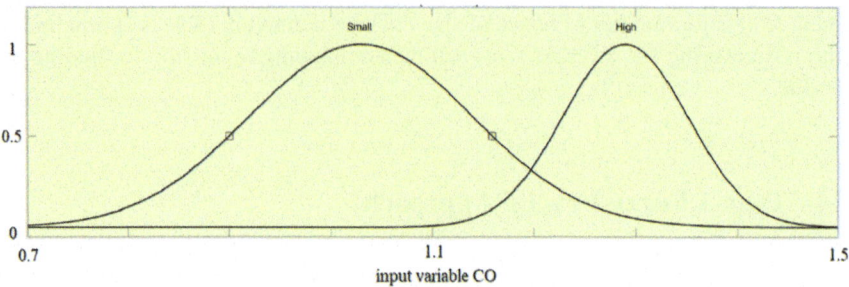

Fig. 3.3 MFs for the FD_{CO} input

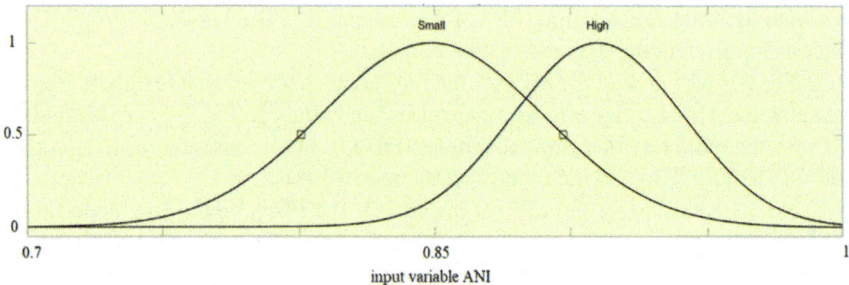

Fig. 3.4 MFs for the FD_{ANI} input

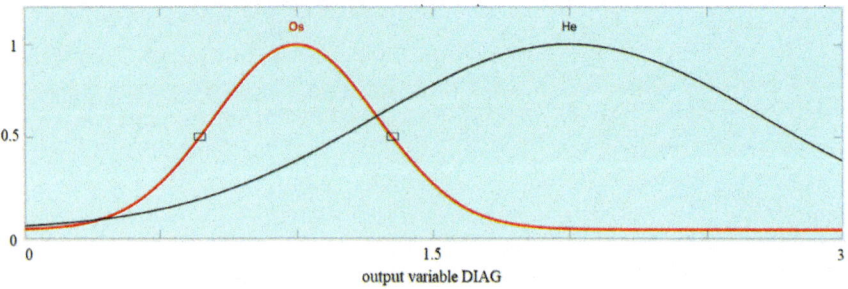

Fig. 3.5 MFs for the DIAG output variable

to be in the range from 0 to 3. All MFs are Gaussian, so we have 2 parameters, center and deviation.

Based on expert knowledge and experimentation with the dataset it is known that an image is classified as Healthy (He) for high values of both FDs, and as Osteoporosis (Os) for low values of both FDs. We also consider that if at least one of the FDs is High then the diagnosis is of osteoporosis, so that more tests could be to a person (making sure of this diagnosis). Then the rules are:

If FD_{CO} is Low and FD_{ANI} is Low Then DIAG is Os

3.5 Proposed Type-3 Approach

Table 3.1 Parameter values of the FD_{C0} input MFs

MFs	Center	Deviation
Low	1.030	0.11
High	1.290	0.06

Table 3.2 Parameter values of the FD_{ANI} input MFs

MFs	Center	Deviation
Low	0.849	0.041
High	0.910	0.031

Table 3.3 Parameter values of the output MFs

MFs	Center	Deviation
Os	1.0	0.30
He	2.0	0.71

If FD_{CO} is High and FD_{ANI} is High Then DIAG is He
If FD_{CO} is High and FD_{ANI} is Low Then DIAG is Os
If FD_{CO} is Low and FD_{ANI} is High Then DIAG is Os

We think of employing other input variables for the system in future work.

3.5 Proposed Type-3 Approach

We now elevate the type-1 system to type-3 by using T3MFs. Now the FD_{CO} input T3MFs are depicted in Fig. 3.6. The FD_{ANI} input MFs are depicted in Fig. 3.7. The output T3MFs are depicted in Fig. 3.8.

In Fig. 3.9 we can find the architecture of the T3FLS in which the fractal dimensions of the roughness and anisotropy enter as inputs, and the diagnosis is the output. The difference with respect to Fig. 3.2 is that now we are employing T3MFs that can handle higher levels of uncertainty, which would in enhance the diagnosis. We also show in Figs. 3.10 and 3.11 a three-dimensional (3-D) view of T3MFs for the inputs and in Fig. 3.12 for the output, respectively.

Table 3.4 summarizes the parameter values for the T3MFs of the FD_{co} input and Table 3.5 for the FD_{ANI} input, respectively. Table 3.6 lists the parameters for the T3MFs of DIAG, which we used in the range from 0 to 3. The T3MFs are scaled Gaussians, so we now have four parameters: center, deviation, lower lag and lower scale [8].

The rules are the same, as now the difference is that the T3FLS employs T3MFs.

Fig. 3.6 Input T3MFs of FD_{CO}

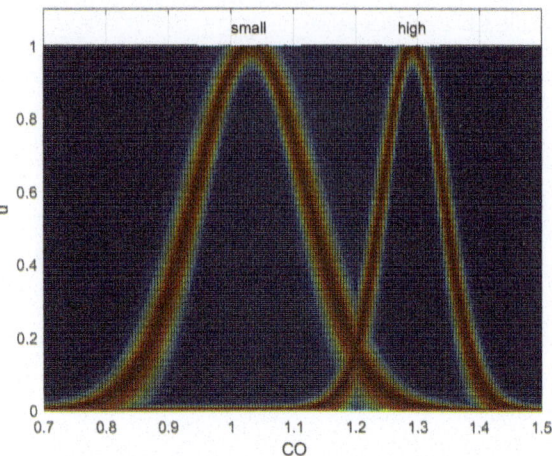

Fig. 3.7 Input T3MFs of FD_{ANI}

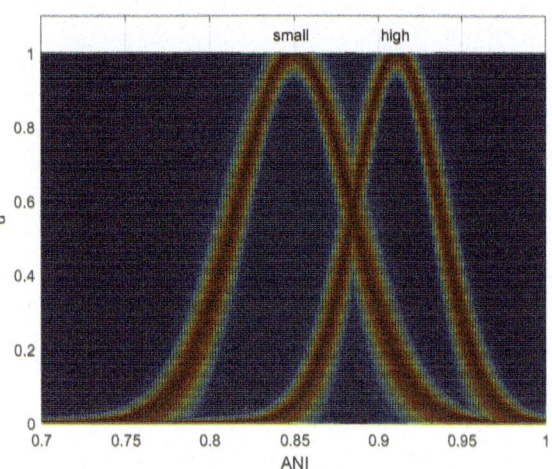

3.6 Application to Osteoporosis Diagnosis

This section is presenting the results of the fuzzy-fractal systems for brain tumor diagnosis. Table 3.7 summarizes a comparison of results, where column one shows the FD values for 12 cases, column two lists the corresponding diagnosis of experts (LG = 1, HG = 2), column three shows the results of type-1, and column four results of type-3. From Table 3.7 we can note that both fuzzy-fractal approaches produce a correct diagnosis in 12 case, however numerically the results of type-3 are closer on average to values of experts.

3.7 Conclusions

Fig. 3.8 Output T3MFs

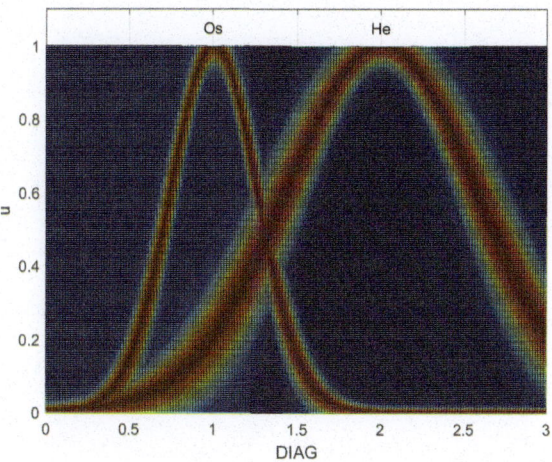

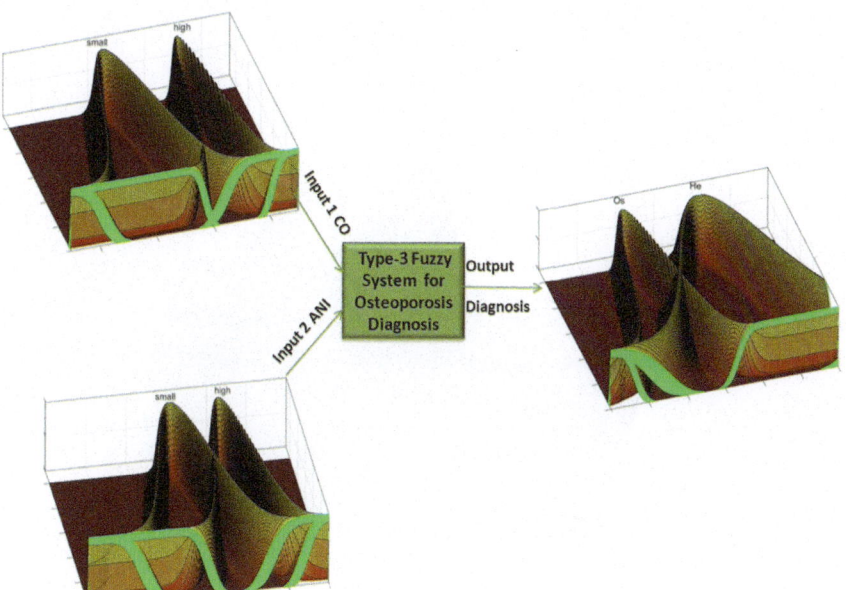

Fig. 3.9 Architecture of the T3FLS for diagnosis

3.7 Conclusions

In this chapter a hybrid of type-3 and FD was employed for brain tumor diagnosis. FD was utilized to characterize the geometrical properties of the bone. A T3FLS was employed to represent the knowledge of experts in diagnosis. Simulation results with a dataset of tumors shows the effectiveness of the proposal. In future work we could

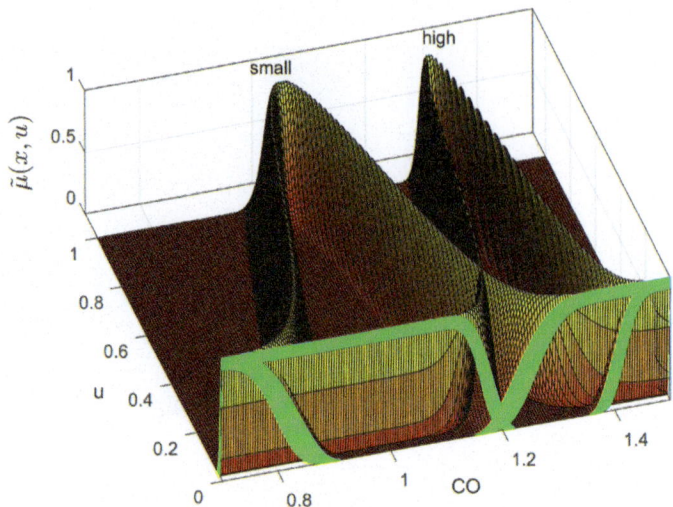

Fig. 3.10 3-D view of FD$_{CO}$ input T3MFs

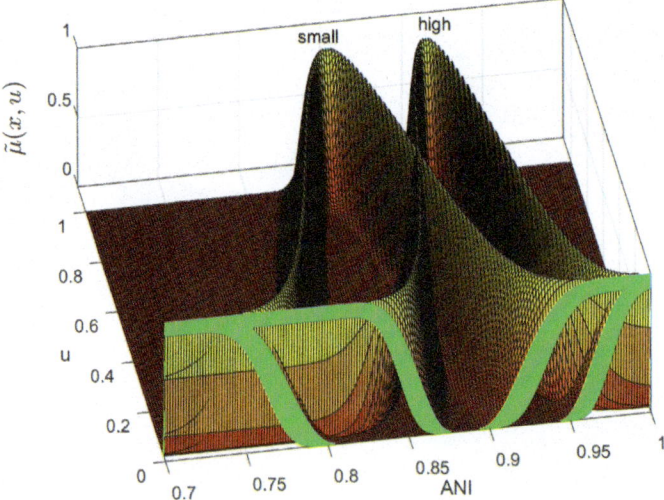

Fig. 3.11 3-D view of FD$_{ANI}$ input T3MFs

employ the diagnosis approach for other medical problems. In addition, we could consider optimizing the parameters of the fuzzy systems by utilizing metaheuristics, such as the ones in [26–29].

3.7 Conclusions

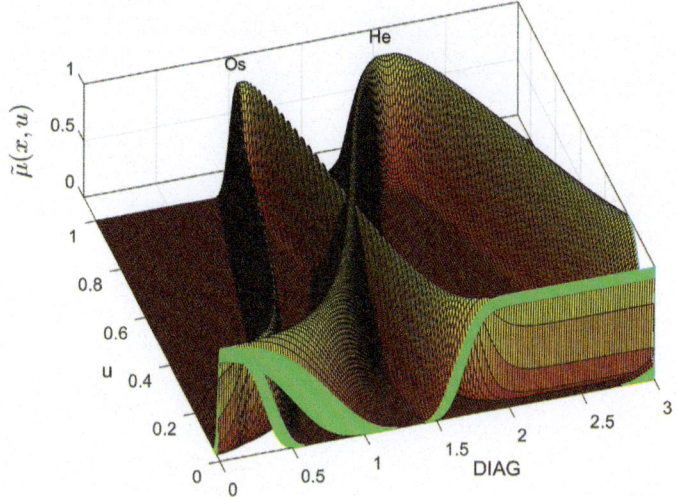

Fig. 3.12 3-D view of output T3MFs

Table 3.4 Parameter values of the input T3MF

MFs	Center	Deviation	Lower scale	Lower lag
Low	1.030	0.11	0.9	0.2
High	1.290	0.06	0.9	0.2

Table 3.5 Parameter values of the input T3MFs

MFs	Center	Deviation	Lower scale	Lower lag
Low	0.849	0.041	0.9	0.2
High	0.910	0.031	0.9	0.2

Table 3.6 Parameter values of the output T3MFs

MFs	Center	Deviation	Lower scale	Lower lag
Os	1.0	0.30	0.9	0.1
He	2.0	0.71	0.9	0.2

Table 3.7 Comparison of fuzzy-fractal approaches for brain tumor diagnosis

CO	ANI	DIAG	Fuzzy type-1 DIAG	Fuzzy type-3 DIAG
1.27	0.90	2 (He)	1.81	1.8524
1.28	0.85	2 (He)	1.76	1.7801
1.01	0.89	1 (Os)	1.02	1.0109
0.95	0.86	1 (Os)	1.03	1.0130
0.93	0.84	1 (Os)	1.04	1.0167
1.29	0.91	2 (He)	1.85	1.8971
1.03	0.84	1 (Os)	1.04	1.0122
1.04	0.92	1 (Os)	1.05	1.0111
1.29	0.83	1 (Os)	1.07	1.0201
1.30	0.90	2 (He)	1.81	1.8529

References

1. L.A. Zadeh, Fuzzy sets. Inf. Control. **8**, 338–353 (1965)
2. L.A. Zadeh, Knowledge representation in fuzzy logic. IEEE Trans. Knowl. Data Eng. **1**, 89 (1989)
3. P. Melin, O. Castillo, Adaptive intelligent control of aircraft systems with a hybrid approach combining neural networks, fuzzy logic and fractal theory. Appl. Soft Comput. 3(4), 353–362 (2003)
4. O. Castillo, P. Melin, Intelligent adaptive model-based control of robotic dynamic systems with a hybrid fuzzy-neural approach. Appl. Soft Comput. 3(4), 363–378 (2003)
5. L Astudillo, O Castillo, P Melin, A Alanis, J Soria, LT Aguilar, Intelligent control of an autonomous mobile robot using type-2 fuzzy logic. Eng. Lett. **13**(3) (2006)
6. J. Urias, D. Hidalgo, P. Melin, O. Castillo, A method for response integration in modular neural networks with type-2 fuzzy logic for biometric systems. Adv. Soft Comput. **41**, 5–15 (2007)
7. R. Sepúlveda, O. Montiel, O. Castillo, P. Melin, Embedding a high speed interval type-2 fuzzy controller for a real plant into an FPGA. Appl. Soft Comput. **12**(3), 988–998 (2012)
8. O. Castillo, J.R. Castro, P. Melin, *Interval Type-3 Fuzzy Systems: Theory and Design* (Springer, Cham, Switzerland, 2022)
9. O. Castillo, P. Melin, Towards interval type-3 intuitionistic fuzzy sets and systems. Mathematics **10**, 4091 (2022)
10. M.W. Tian, S.R. Yan, A. Mohammadzadeh, J. Tavoosi, S. Mobayen, R. Safdar, W. Assawinchaichote, M.T. Vu, A. Zhilenkov, Stability of interval type-3 fuzzy controllers for autonomous vehicles. Mathematics **9**(21), 2742 (2021)
11. J.T. Rickard, J. Aisbett, G. Gibbon, Fuzzy subsethood for fuzzy sets of type-2 and generalized type-n. IEEE Trans. Fuzzy Syst. **17**(1), 50–60 (2009)
12. A. Mohammadzadeh, M.H. Sabzalian, W. Zhang, An interval type-3 fuzzy system and a new online fractional-order learning algorithm: theory and practice. IEEE Trans. Fuzzy Syst. **28**(9), 1940–1950 (2020)
13. Z. Liu, A. Mohammadzadeh, H. Turabieh, M. Mafarja, S.S. Band, A. Mosavi, A new online learned interval type-3 fuzzy control system for solar energy management systems. IEEE Access **9**, 10498–10508 (2021)
14. D. Singh, N.K. Verma, A.K. Ghosh, A.K. Malagaudanavar, An approach towards the design of interval type-3 ts fuzzy system. IEEE Trans. Fuzzy Syst. (2021)
15. J.H. Wang, J. Tavoosi, A. Mohammadzadeh, S. Mobayen, J.H. Asad, W. Assawinchaichote, P. Skruch, Non-singleton type-3 fuzzy approach for flowmeter fault detection: experimental study in a gas industry. Sensors **21**(21), 7419 (2021)

16. K.A. Alattas, A. Mohammadzadeh, S. Mobayen, A.A. Aly, B.F. Felemban, A new data-driven control system for memss gyroscopes: dynamics estimation by type-3 fuzzy systems. Micromachines **12**(11), 1390 (2021)
17. Y. Cao, A. Raise, A. Mohammadzadeh, S. Rathinasamy, S.S. Band, A. Mosavi, Deep learned recurrent type-3 fuzzy system: application for renewable energy modeling/prediction. Energy Rep. **7**, 8115–8127 (2021)
18. A. Chan, J.A. Tuszynski, Automatic prediction of tumour malignancy in breast cancer with fractal dimension. R. Soc. Open Sci. **3**(12), 160558 (2016)
19. Y.D. Zhang, X.Q. Chen, T.M. Zhan, Z.Q. Jiao, Y. Sun, Z.M. Chen, S.H. Wang, Fractal dimension estimation for developing pathological brain detection system based on Minkowski-Bouligand method. IEEE Access **4**, 5937–5947 (2016)
20. O. Alpar, A mathematical fuzzy fusion framework for whole tumor segmentation in multimodal MRI using Nakagami imaging. Expert Syst. Appl. **216**, 119462 (2023)
21. O. Alpar, R. Dolezal, P. Ryska, O. Krejcar, Nakagami-Fuzzy imaging framework for precise lesion segmentation in MRI. Pattern Recogn. **128**, 108675 (2022)
22. O. Alpar, O. Krejcar, Whole tumor area estimation in incremental brain MRI using dilation and erosion-based binary morphing, in *International Work-Conference on Bioinformatics and Biomedical Engineering* (Springer Nature, Cham, 2023), pp. 131–142
23. O. Alpar, O. Krejcar, Three-dimensional representation and visualization of high-grade and low-grade glioma by Nakagami imaging, in *International Work-Conference on Bioinformatics and Biomedical Engineering* (Springer Nature, Cham, 2023), pp. 143–154
24. S.M. Nazia Fathima, R. Tamilselvi, M. Parisa Beham, XSITRAY: a database for the detection of osteoporosis conditions. Biomed. Pharmacol. J. **12**(1), 267–271 (2019)
25. T. Loussot, R. Harba, G. Jacquet, C.L. Benhamou, E. Lespessailles, A. Julien, An oriented fractal analysis for the characterization of texture: application to bone radiographs, in *1996 8th European Signal Processing Conference (EUSIPCO 1996)* (Trieste, Italy, 1996), pp. 1–4
26. B. González, F. Valdez, P. Melin, G. Prado-Arechiga, Fuzzy logic in the gravitational search algorithm for the optimization of modular neural networks in pattern recognition. Expert Syst. Appl. **42**(14), 5839–5847 (2015)
27. L. Amador-Angulo, O. Mendoza, J.R. Castro, A. Rodriguez-Diaz, P. Melin, O. Castillo, Fuzzy sets in dynamic adaptation of parameters of a bee colony optimization for controlling the trajectory of an autonomous mobile robot. Sensors **16**(9), 1458 (2016)
28. O. Castillo, E. Lizarraga, J. Soria, P. Melin, F. Valdez, New approach using ant colony optimization with ant set partition for fuzzy control design applied to the ball and beam system. Inf. Sci. **294**, 203–215 (2015)
29. R. Martínez-Soto, O. Castillo, L.T. Aguilar, Type-1 and type-2 fuzzy logic controller design using a hybrid PSO–GA optimization method. Inf. Sci. **285**, 35–49 (2014)

Chapter 4
A Type-3 Fuzzy-Fractal Approach for Leukemia Cancer Diagnosis

In this chapter a combination of type-3 fuzzy logic and the fractal dimension (FD) is employed for leukemia cancer diagnosis. Leukemia is cancer of blood. FD is utilized to characterize the geometrical properties in the blood cells of a patient. A type-3 fuzzy system is employed to represent the knowledge of experts in diagnosis. In this case, type-3 helps in representing the uncertainty in making a medical diagnosis. Results with a dataset of cases show the effectiveness of the approach.

4.1 Introduction

Now it is known that fuzzy systems generate remarkable applications in many areas, like: control, forecasting and plant monitoring. Originally, fuzzy sets (now called type-1) were put forward by Lotfi Zadeh in 1965 [1, 2]. Later fuzzy logic and fuzzy systems were also proposed by Zadeh, and many applications follow, mainly in control [3–5]. Fuzzy Logic has evolved from the original works of Zadeh with type-1 fuzzy theory [1], to later the developments of type-2 and its applications [6–8], to now where type-3 theory [9] and its applications are emerging [10–16].

In addition, type-3 has been employed, with relative success, to handle uncertainty from sensor measurements, like in control [17–20]. There have been some previous works applying the fractal dimension (FD) to diagnosis in diverse medical problems, as in [21–23]. Also, there exist some works using fuzzy logic in medical diagnosis, like in [24, 25]. For the case of Leukemia diagnosis there exists previous works using FD, like in [26]. However, unlike previous works, in this paper we are offering for the first time a type-3 fuzzy-fractal approach for leukemia cancer diagnosis.

The rest of the chapter is: Sect. 4.2 is summarizing type-3 terminology, in Sect. 4.3 the fractal dimension concept is offered, in Sect. 4.4 the type-1 fuzzy fractal diagnosis is delineated, in Sect. 4.5 we are offering the type-3 fuzzy fractal diagnosis approach, in Sect. 4.6 results are presented, and in Sect. 4.7 conclusions are outlined.

4.2 Type-3 Fuzzy Logic

The initial concept is given below.

Definition 4.1 A type-3 fuzzy set (T3 FS) [9], denoted by $A^{(3)}$, is represented by the membership function (MF) of $A^{(3)}$, in the Cartesian product $X \times [0, 1] \times [0, 1]$ in $[0, 1]$, where X is the primary variable universe of $A^{(3)}$, x. The MF of $\mu_{A^{(3)}}$ is a type-3 MF (T3 MF):

$$\mu_{A^{(3)}} : X \times [0, 1] \times [0, 1] \to [0, 1]$$
$$A^{(3)} = \{(x, u(x), v(x, u), \mu_{A^{(3)}}(x, u, v)) | x \in X, u \in U \subseteq [0, 1], v \in V \subseteq [0, 1]\} \quad (4.1)$$

where U is universe for secondary variable u and V is universe for tertiary variable v. When T3MFs are employed in the fuzzy rules then we have a type-3 fuzzy logic system (T3FLS). The inference process and type reduction calculations are similar to type-2, but more details can be found in [9].

4.3 Fractal Dimension

Recently, significant progress has been achieved in comprehending the complexity of an object by employing fractal constructs [21]. For example, time series and images in medical problems exhibit properties suggesting a fractal structure [22, 23]. The FD is postulated as:

$$d = \lim_{r \to 0} [lnN(r)]/[ln(1/r)] \quad (4.2)$$

where $N(r)$ is for number of boxes of size r. The d defined in (4.2) is approximated with box covering for several r sizes and then employing regression for estimating the d value. For the particular case of Leukemia, in [26] it was found that the Mean fractal dimension of lymphoma cell was 1.367 ± 0.0011 in healthy subjects and 1.398 ± 0.0016 in cancer patients. Also, this difference between FDs of healthy and cancer cells was significant. Based on this information, we propose to consider the fuzzy concepts for handling uncertainty in the decision process for diagnosis.

4.4 Type-1 Fuzzy-Fractal Approach

We are now presenting a fuzzy fractal approach for leukemia cancer diagnosis. In Fig. 4.1 we can notice the structure of the system in which the FD enters as input, and the diagnosis is the output.

4.4 Type-1 Fuzzy-Fractal Approach

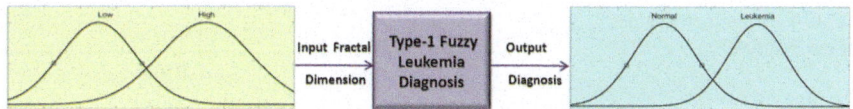

Fig. 4.1 Structure of the fuzzy fractal approach for Leukemia diagnosis

In Fig. 4.2 we can find the MFs for the FD, which are Low and High.

In Fig. 4.3 we can find the MFs for the output, which is the diagnosis (DIAG). In this case, there are two MFs, one for Healthy (He) and the other for Leukemia (Le).

Table 4.1 summarizes the parameter values for the MFs of the FD input. The range for the FD is from 1.34 to 1.42, as experimentally it has been found that the FDs of the blood cells are in this range. Table 4.2 presents the parameters for the MFs of DIAG, which we consider to be in the range from 0 to 3. All the MFs are Gaussian, so we only have two parameters, center and deviation.

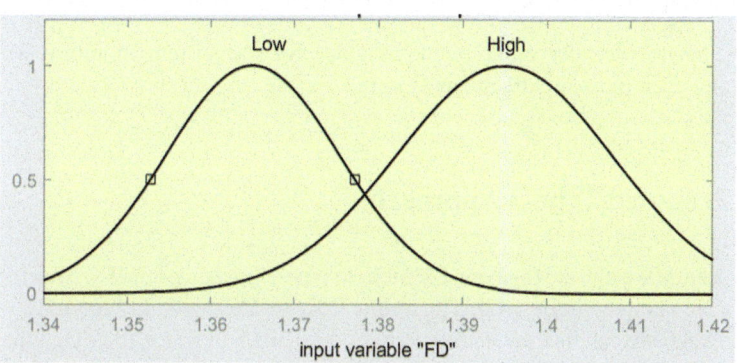

Fig. 4.2 MFs for the FD input

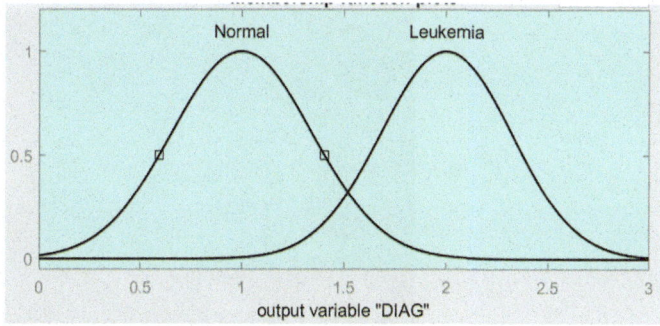

Fig. 4.3 MFs for the DIAG output variable

Table 4.1 Parameter values of the input MFs

MFs	Center	Deviation
Low	1.367	0.01041
High	1.398	0.01285

Table 4.2 Parameter values of the output MFs

MFs	Center	Deviation
LG	1.0	0.3452
HG	2.0	0.3183

Based on expert knowledge and experimentation with a blood cell dataset it is known that a subject is classified as Healthy (He) for low values of FD, and a subject is classified as with Leukemia (Le) for high values of FD. Then the rules are:

If FD is Low Then DIAG is He
If FD is High Then DIAG is Le

We plan to consider other input variables for the system in future work for enhancing even more the diagnosis system.

4.5 Proposed Type-3 Approach

In this section we elevate the type-1 system to type-3 by using T3MFs. Now the input T3MFs are depicted in Fig. 4.4. The output T3MFs are shown in Fig. 4.5.

In Fig. 4.6 we can find the architecture of the T3FLS in which the FD of the blood cells enters as input, and the diagnosis is the output.

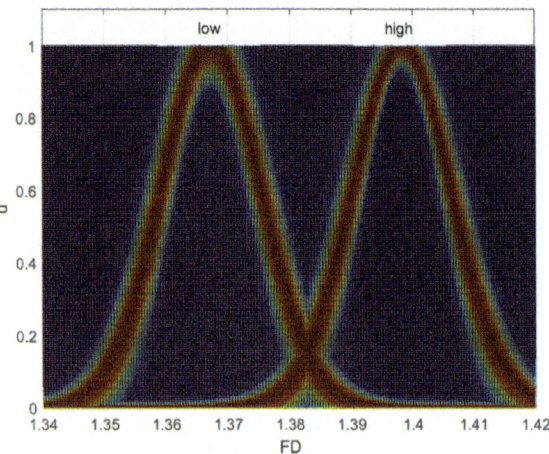

Fig. 4.4 Input T3MFs

4.5 Proposed Type-3 Approach

Fig. 4.5 Output T3MFs

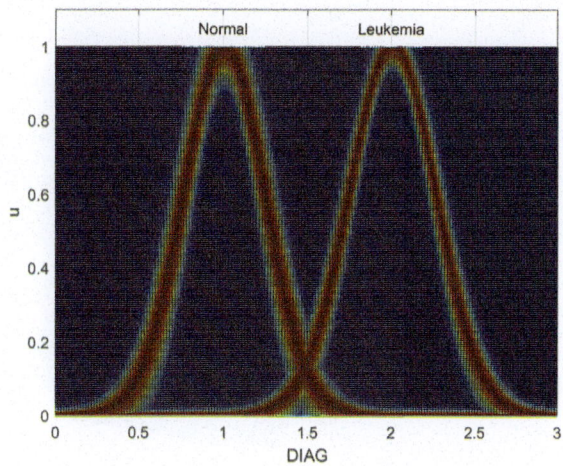

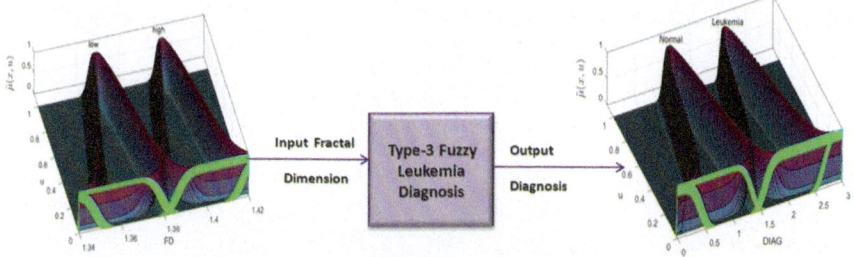

Fig. 4.6 Architecture of the T3FLS for Leukemia diagnosis

The difference with respect to Fig. 4.1 is that now we are employing T3MFs that can handle higher levels of uncertainty, which would in enhance the diagnosis. We also show in Figs. 4.7 and 4.8 a three-dimensional (3-D) view of T3MFs for the input and output, respectively.

Table 4.3 summarizes the parameter values for the T3MFs of the FD input. Table 4.4 lists the parameters for the T3MFs of DIAG, which we used in the range from 0 to 3. The T3MFs are scaled Gaussians, so we now have four parameters: center, deviation, lower lag and lower scale [9].

The rules are the same, as now the difference is that the T3FLS employs T3MFs.

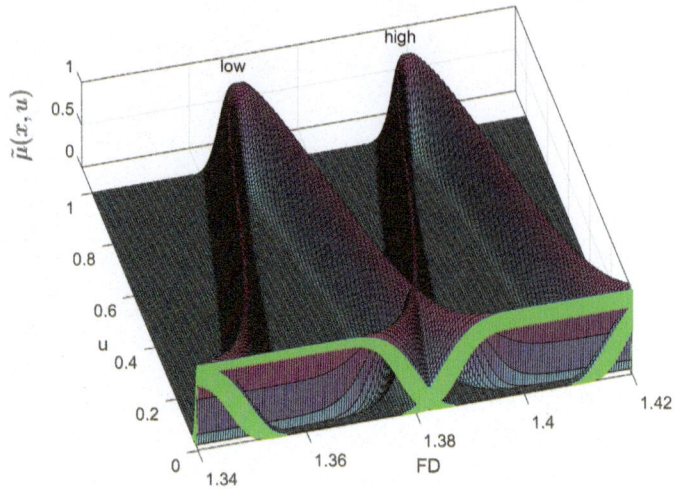

Fig. 4.7 3-D view of input T3MFs

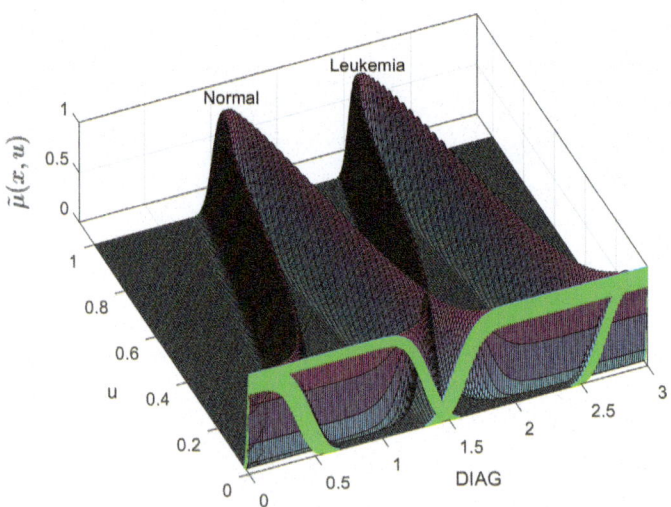

Fig. 4.8 3-D view of output T3MFs

Table 4.3 Parameter values of the input T3MFs

MFs	Center	Deviation	Lower scale	Lower lag
Low	1.367	0.010	0.8	0.2
High	1.398	0.010	0.9	0.2

Table 4.4 Parameter values of the output T3MFs

MFs	Center	Deviation	Lower scale	Lower lag
He	1.0	0.3	0.8	0.2
Le	2.0	0.3	0.9	0.1

Table 4.5 Comparison of fuzzy-fractal approaches for leukemia diagnosis

FD	DIAG	Type-1 Fuzzy DIAG	Type-3 Fuzzy DIAG
1.365	1 (Normal)	1.0900	1.0095
1.397	2 (Leukemia)	1.9800	1.9897
1.371	1 (Normal)	1.2100	1.0142
1.401	2 (Leukemia)	1.9900	1.9905
1.368	1 (Normal)	1.1400	1.0104
1.399	2 (Leukemia)	1.9900	1.9904
1.386	2 (Leukemia)	1.7900	1.8352
1.389	2 (Leukemia)	1.8800	1.9463
1.359	1 (Normal)	1.0300	1.0116
1.361	1 (Normal)	1.0500	1.0102
1.379	1 (Normal)	1.5000	1.1926
1.395	2 (Leukemia)	1.9700	1.9881

4.6 Application to Leukemia Diagnosis

In this section we are presenting the results of the fuzzy-fractal approaches for leukemia diagnosis. Table 4.5 summarizes a comparison of results, where column one shows the FD values for 12 cases, column two lists the corresponding diagnosis of experts (He $= 1$, Le $= 2$), column three shows the results of type-1, and column four the results of type-3.

From Table 4.5 we can note that both fuzzy-fractal approaches produce a correct diagnosis in 12 cases, however numerically the results of type-3 are closer on average to values of experts.

4.7 Conclusions

In this chapter a hybrid of type-3 and FD was employed for leukemia diagnosis. FD was utilized to characterize the geometrical properties of the blood cells. A type-3 fuzzy system was employed to represent the knowledge of experts in diagnosis. Simulation results with a dataset of samples shows the effectiveness of the approach. In future work we will apply the approach to the diagnosis of other medical problems.

References

1. L.A. Zadeh, Fuzzy sets. Inf. Control **8**, 338–353 (1965)
2. L.A. Zadeh, Knowledge representation in fuzzy logic. IEEE Trans. Knowl. Data Eng. **1**, 89 (1989)
3. P. Melin, O. Castillo, Adaptive intelligent control of aircraft systems with a hybrid approach combining neural networks, fuzzy logic and fractal theory. Appl. Soft Comput. **3**(4), 353–362 (2003)
4. O. Castillo, P. Melin, Intelligent adaptive model-based control of robotic dynamic systems with a hybrid fuzzy-neural approach. Appl. Soft Comput. **3**(4), 363–378 (2003)
5. O. Castillo, F. Kutlu, O. Atan, Intuitionistic fuzzy control of twin rotor multiple input multiple output systems. J. Intell. Fuzzy Syst. **38**(1), 821–833 (2020)
6. J.M. Mendel, R.I. Bob John, Type-2 fuzzy sets made simple. IEEE Trans. Fuzzy Syst. **10**(2), 117–127 (2002)
7. P. Melin, O. Castillo, A new method for adaptive control of non-linear plants using type-2 fuzzy logic and neural networks. Int. J. Gen Syst **33**(2–3), 289–304 (2004)
8. J.R. Castro, O. Castillo, P. Melin, A. Rodríguez-Díaz, (2008). Building fuzzy inference systems with a new interval type-2 fuzzy logic toolbox, in *Transactions on computational science I*. Lecture Notes in Computer Science, vol 4750 (Springer, Berlin, Heidelberg, 2008), pp. 104–114. https://doi.org/10.1007/978-3-540-79299-4_5
9. O. Castillo, J.R. Castro, P. Melin, *Interval Type-3 Fuzzy Systems: Theory and Design* (Springer, Cham, Switzerland, 2022)
10. O. Castillo, P. Melin, Towards interval type-3 intuitionistic fuzzy sets and systems. Mathematics **10**, 4091 (2022). https://doi.org/10.3390/math10214091
11. S.N. Qasem, A. Ahmadian, A. Mohammadzadeh, S. Rathinasamy, B. Pahlevanzadeh, A type-3 logic fuzzy system: Optimized by a correntropy based Kalman filter with adaptive fuzzy kernel size. Inform. Sci. **572**, 424–443 (2021)
12. A. Mohammadzadeh, M.H. Sabzalian, W. Zhang, An interval type-3 fuzzy system and a new online fractional-order learning algorithm: theory and practice. IEEE Trans. Fuzzy Syst. **28**(9), 1940–1950 (2020)
13. Z. Liu, A. Mohammadzadeh, H. Turabieh, M. Mafarja, S.S. Band, A. Mosavi, A new online learned interval type-3 fuzzy control system for solar energy management systems. IEEE Access **9**, 10498–10508 (2021)
14. Singh, D., Verma, N. K., Ghosh, A. K., & Malagaudanavar, A. K. (2021). An Approach Towards the Design of Interval Type-3 TS Fuzzy System. *IEEE Transactions on Fuzzy Systems*.
15. J.H. Wang, J. Tavoosi, A. Mohammadzadeh, S. Mobayen, J.H. Asad, W. Assawinchaichote, P. Skruch, Non-singleton type-3 fuzzy approach for flowmeter fault detection: experimental study in a gas industry. Sensors **21**(21), 7419 (2021)
16. K.A. Alattas, A. Mohammadzadeh, S. Mobayen, A.A. Aly, B.F. Felemban, A new data-driven control system for MEMSs gyroscopes: dynamics estimation by type-3 fuzzy systems. Micromachines **12**(11), 1390 (2021)
17. N. Nabipour, S.N. Qasem, K. Jermsittiparsert, Type-3 fuzzy voltage management in PV/hydrogen fuel cell/battery hybrid systems. Int. J. Hydrog. Energy **45**(56), 32478–33249 (2020)
18. A. Taghieh, A.A. Aly, B.F. Felemban, A. Althobaiti, A. Mohammadzadeh, A. Bartoszewicz, A hybrid predictive type-3 fuzzy control for time-delay multi-agent systems. Electronics **11**(1), 63 (2022)
19. R.H. Vafaie, A. Mohammadzadeh, M. Piran, A new type-3 fuzzy predictive controller for MEMS gyroscopes. Nonlinear Dyn. **106**(1), 381–403 (2021)
20. M.W. Tian, S.R. Yan, A. Mohammadzadeh, J. Tavoosi, S. Mobayen, R. Safdar, W. Assawinchaichote, M.T. Vu, A. Zhilenkov, Stability of interval type-3 fuzzy controllers for autonomous vehicles. Mathematics **9**(21), 2742 (2021)
21. A. Chan, J.A. Tuszynski, Automatic prediction of tumour malignancy in breast cancer with fractal dimension. R. Soc. Open Sci. **3**(12), 160558 (2016)

References

22. Y.D. Zhang, X.Q. Chen, T.M. Zhan, Z.Q. Jiao, Y. Sun, Z.M. Chen, S.H. Wang, Fractal dimension estimation for developing pathological brain detection system based on Minkowski-Bouligand method. IEEE Access **4**, 5937–5947 (2016)
23. T. Loussot, R. Harba, G. Jacquet, C.L. Benhamou, E. Lespessailles, A. Julien, An oriented fractal analysis for the characterization of texture: application to bone radiographs, in *1996 8th European Signal Processing Conference (EUSIPCO 1996)* (Trieste, Italy, 1996), pp. 1–4
24. O. Alpar, A mathematical fuzzy fusion framework for whole tumor segmentation in multimodal MRI using Nakagami imaging. Expert Syst. Appl. **216**, 119462 (2023)
25. O. Alpar, R. Dolezal, P. Ryska, O. Krejcar, Nakagami-fuzzy imaging framework for precise lesion segmentation in MRI. Pattern Recogn. **128**, 108675 (2022)
26. T. Yousefzadeh Hassanluie, M. Reza Rezaie, Z. Rostami, Diagnosis of B-CLL leukemia using fractal dimension. J. Kerman Univ. Med. Sci. **24**(3), 229–236 (2017)

Chapter 5
A Type-3 Fuzzy-Fractal Approach for Diagnosis of Vascular Diseases Based on Cardiac Vessels

In this chapter a combination of type-3 fuzzy logic system (T3FLS) and the fractal dimension (FD) is employed for diagnosis of vascular diseases. FD is utilized to characterize the geometrical properties of the cardiac vessels. A T3FLS is employed to represent the knowledge of experts in vascular diagnosis. In this case, the T3FLS helps in representing the uncertainty in making a medical diagnosis. Simulation results with a dataset of tumors shows the accuracy of the proposal.

5.1 Introduction

Fuzzy systems can achieve remarkable applications in many areas, like: forecasting and monitoring. Originally, fuzzy sets were offered by Zadeh in 1965 [1, 2]. Later fuzzy systems were also postulated, and many applications follow, like in [3, 4]. Fuzzy Logic has evolved from the initial studies with type-1 fuzzy theory [1], to later the theory of type-2 [5–7], to now where type-3 is appearing [8–10]. Now even type-n has been mentioned [11]. In addition, type-3 has been employed to handle uncertainty from measurements, like in control [12–17]. There have been some previous works applying the fractal dimension (FD) to diagnosis in diverse medical problems, as in [18, 19]. Also, there exist some works using fuzzy logic in medical diagnosis, like in [20, 21].

The stenosis of a coronary artery is the beginning of the coronary artery disease which in time will create an obstruction of the blood flow. Furthermore, the tissue which was fed by that vessel will not receive enough nutriments and will die creating a necrosis into the heart. This process will perturb the heart performance. The diagnosis of this disease implies an echocardiography, computer tomography and/or angiography. Unlike previous works, in this paper we are offering for the first time a type-3

© The Author(s), under exclusive license to Springer Nature Switzerland AG 2025
P. Melin and O. Castillo, *Type-3 Fuzzy Logic and Fractal Theory for Medical Diagnosis*, SpringerBriefs in Computational Intelligence,
https://doi.org/10.1007/978-3-031-81655-0_5

fuzzy-fractal hybrid for diagnosis of vascular diseases based on cardiac vessels. We employ values from a dataset of x-ray images from cardiac vessels [22–25] for testing our approach.

The rest of the document is: Sect. 5.2 is summarizing type-3 terminology, in Sect. 5.3 the fractal dimension concept is offered, in Sect. 5.4 the type-1 fuzzy fractal diagnosis is delineated, in Sect. 5.5 we are offering the type-3 fuzzy fractal diagnosis approach, in Sect. 5.6 results are presented, and in Sect. 5.7 conclusions are offered.

5.2 Type-3 Fuzzy Theory

We first postulate the concepts.

Definition 5.1 A type-3 fuzzy set (T3 FS) [8], written as $A^{(3)}$, is represented by the membership function (MF) of $A^{(3)}$, in the Cartesian product $X \times [0, 1] \times [0, 1]$ in $[0, 1]$, where X is the primary variable universe of $A^{(3)}$, x. The MF of $\mu_{A^{(3)}}$ is a type-3 MF (T3 MF):

$$\mu_{A^{(3)}} : X \times [0, 1] \times [0, 1] \to [0, 1]$$
$$A^{(3)} = \{(x, u(x), v(x, u), \mu_{A^{(3)}}(x, u, v)) | x \in X, u \in U \subseteq [0, 1], v \in V \subseteq [0, 1]\} \tag{5.1}$$

where U is universe for secondary variable u and V is universe for tertiary variable v. When T3MFs are employed in the fuzzy rules then we have a type-3 fuzzy logic system (T3FLS).

5.3 Fractal Dimension

Recently, remarkable advances have been achieved in comprehending the complexity of an object by employing fractal theory [18, 19]. For example, time series in finance suggest fractal structure. There are several definitions of FD, but here we utilize one of the simplest ones. The FD is postulated as:

$$d = \lim_{r \to 0} [lnN(r)]/[ln(1/r)] \tag{5.2}$$

where $N(r)$ is for number of boxes of size r. The d defined in (5.2) is approximated with box covering for several r sizes and then employing regression for estimating the d value.

5.4 Type-1 Fuzzy-Fractal Approach

For the cardiac angiograms, two different situations are considered, one for healthy and one for stenosis vessels. The FDs for the two above mentioned images are computed and the mean FD for the stenosis cases is 1.8586 and for the normal cases is 1.8265. It can be observed an FD increase for stenosis.

5.4 Type-1 Fuzzy-Fractal Approach

We are presenting a fuzzy fractal approach for diagnosis of vascular diseases. In Fig. 5.1 we can notice the fuzzy system structure, in which the fractal dimension (FD) of the cardiac vessels [25] enters as input, and the diagnosis is the output.

In Fig. 5.2 we can find the MFs for the FD, which are Low and High. In Fig. 5.3 we can find the MFs for the output, which is the diagnosis (DIAG). In this case, there are two MFs, one for Healthy (He) and the other for Stenosis (St), which is the narrowing of the vessels that causes problems.

Table 5.1 summarizes the MF parameters for the FD input. The range for the FD is from 1.81 to 1.87, as experimentally it has been found that the FDs of the cardiac vessels are in this range. Table 5.2 presents the parameters for the MFs of DIAG, which we consider to be in the range from 0 to 3. All MFs are Gaussian, so we have 2 parameters, center and deviation.

Based on expert knowledge and experimentation with the dataset it is known that an image is classified as Healthy (normal) for low values of FD, and is classified as Stenosis (dangerous) for high values of FD. Then the rules are:

If FD is Low Then DIAG is He

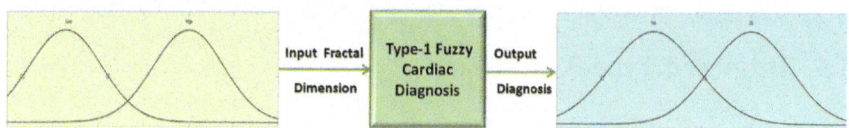

Fig. 5.1 Structure of the fuzzy fractal approach for cardiac vascular diagnosis

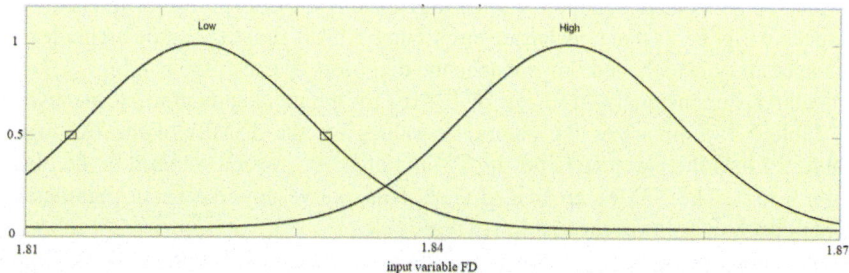

Fig. 5.2 MFs for the FD input

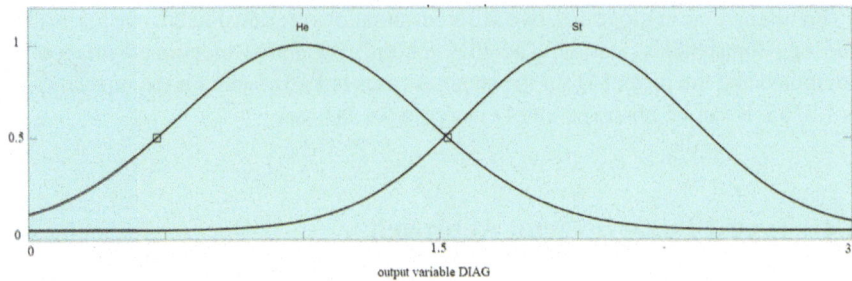

Fig. 5.3 MFs for the DIAG output variable

Table 5.1 Parameter values of the input MFs

MFs	Center	Deviation
Low	1.8241	0.0055
High	1.8563	0.0057

Table 5.2 Parameter values of the output MFs

MFs	Center	Deviation
He	1.0	0.30
St	2.0	0.30

If FD is High Then DIAG is St

We think of employing other input variables for the fuzzy system in future work.

5.5 Proposed Type-3 Approach

We now elevate the type-1 fuzzy system to type-3 by using T3MFs. Now the input T3MFs are depicted in Fig. 5.4. The output T3MFs are shown in Fig. 5.5.

In Fig. 5.6 we can find the architecture of the T3FLS in which the FD of the cardiac vessels enters as input, and the diagnosis is the output. The difference with respect to Fig. 5.1 is that now we are employing T3MFs that can handle higher levels of uncertainty, which would in enhance the diagnosis. We also show in Figs. 5.7 and 5.8 a three-dimensional (3-D) view of T3MFs for the input and output, respectively.

Table 5.3 summarizes the parameter values for the T3MFs of the FD input. Table 5.4 lists the parameters for the T3MFs of DIAG, which we used in the range from 0 to 3. The T3MFs are scaled Gaussians, so we now have four parameters: center, deviation, lower lag and lower scale [8].

The rules are the same, as now the difference is that the T3FLS employs T3MFs.

5.5 Proposed Type-3 Approach

Fig. 5.4 Input T3MFs

Fig. 5.5 Output T3MFs

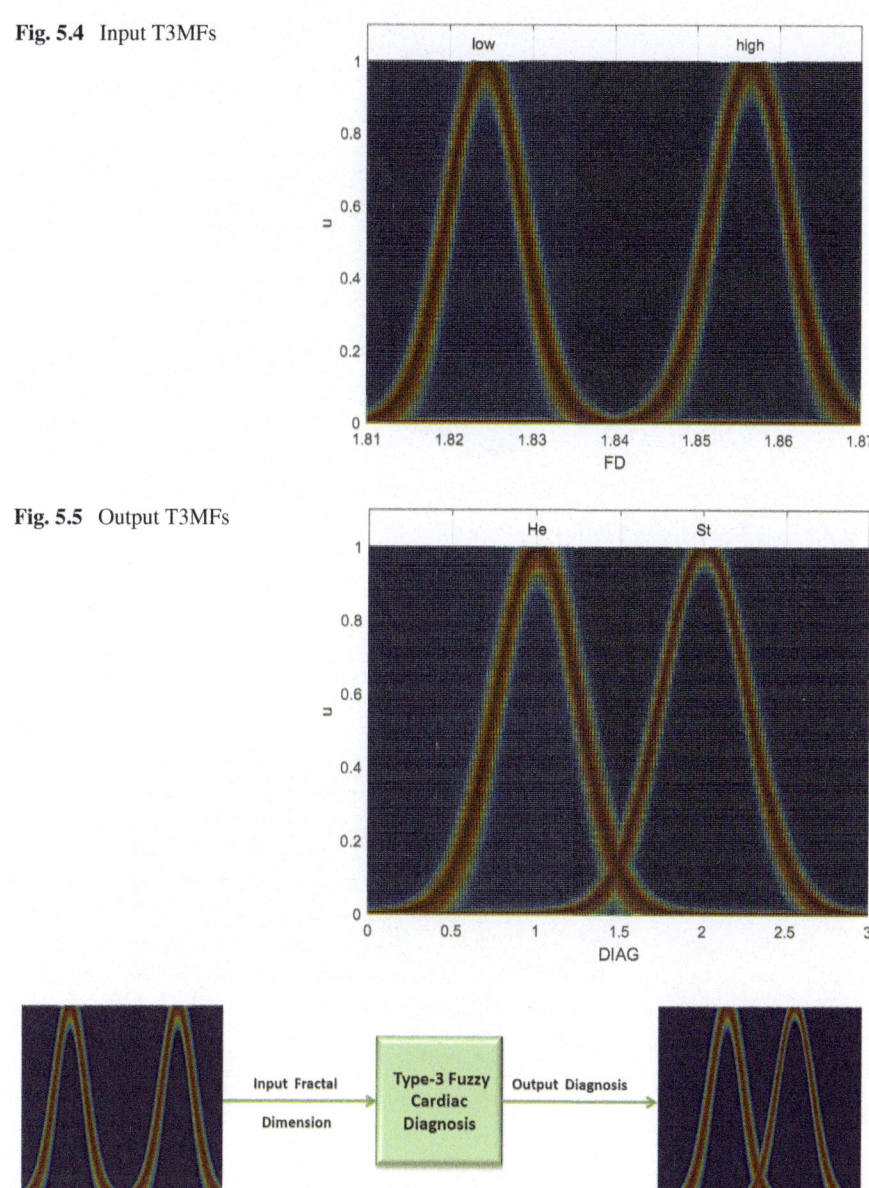

Fig. 5.6 Architecture of the T3FLS for cardiac vascular diagnosis

42 5 A Type-3 Fuzzy-Fractal Approach for Diagnosis of Vascular Diseases …

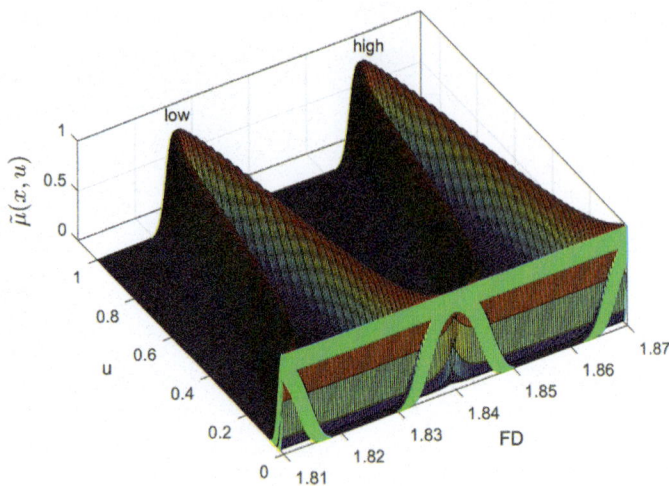

Fig. 5.7 3-D view of input T3MFs

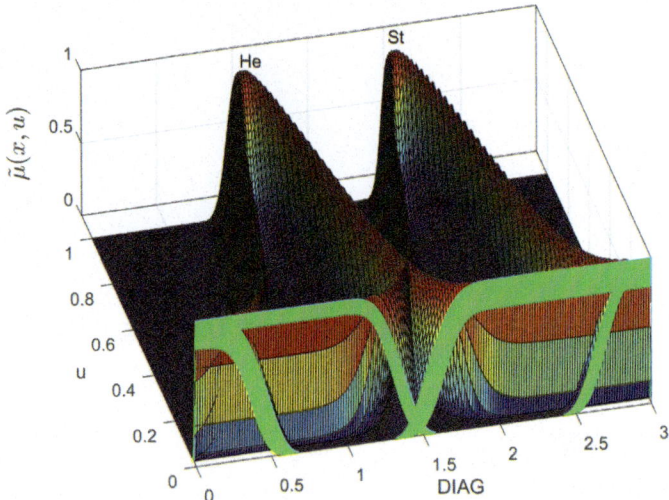

Fig. 5.8 3-D view of output T3MFs

Table 5.3 Parameter values of the input T3MFs

MFs	Center	Deviation	Lower scale	Lower lag
Low	1.8241	0.0055	0.8	0.2
High	1.8563	0.0057	0.8	0.2

Table 5.4 Parameter values of the output T3MFs

MFs	Center	Deviation	Lower scale	Lower lag
He	1.0	0.3	0.8	0.2
St	2.0	0.3	0.9	0.1

Table 5.5 Comparison of fuzzy-fractal approaches for diagnosis of vascular diseases

FD	DIAG	Type-1 fuzzy DIAG	Type-3 fuzzy DIAG
1.810	He	1.03 (He)	1.2353 (He)
1.820	He	1.01 (He)	1.0111 (He)
1.825	He	1.02 (He)	1.0092 (He)
1.830	He	1.06 (He)	1.0152 (He)
1.835	He	1.32 (He)	1.0875 (He)
1.840	St	1.80 (St)	1.5357 (St)
1.850	St	1.99 (St)	1.9841 (St)
1.855	St	1.99 (St)	1.9907 (St)
1.860	St	1.98 (St)	1.9895 (St)
1.865	St	1.96 (St)	1.9686 (HG)

5.6 Application to Brain Tumor Diagnosis

This section is presenting the results of the fuzzy-fractal systems for cardiovascular disease diagnosis. Table 5.5 summarizes a comparison of results, where column one shows the FD values for 10 cases, column two lists the corresponding diagnosis of experts (He = 1, St = 2), column three shows the results of type-1, and column four results of type-3. From Table 5.5 we can note that both fuzzy-fractal approaches produce a correct diagnosis in the 10 cases, however numerically the results of type-3 are closer on average to values of the experts.

5.7 Conclusions

In this chapter a hybrid of type-3 and FD was employed for the diagnosis of vascular diseases. FD was utilized to characterize the geometrical properties of the cardiac vessels. A type-3 system was employed to represent the knowledge of experts in diagnosis. Simulation results with a dataset of x-ray images shows the effectiveness of the proposal. In future work we could employ the diagnosis approach for other medical problems. In addition, we envision optimizing structure and parameters of the fuzzy system with the idea of enhancing even more the results, as have been done in works like [26–30].

References

1. L.A. Zadeh, Fuzzy sets. Inf. Control **8**, 338–353 (1965)
2. L.A. Zadeh, Knowledge representation in Fuzzy Logic. IEEE Trans. Knowl. Data Eng. **1**, 89 (1989)
3. P. Melin, O. Castillo, Adaptive intelligent control of aircraft systems with a hybrid approach combining neural networks, fuzzy logic and fractal theory. Appl. Soft Comput. **3**(4), 353–362 (2003)
4. O. Castillo, P. Melin, Intelligent adaptive model-based control of robotic dynamic systems with a hybrid fuzzy-neural approach. Appl. Soft Comput. **3**(4), 363–378 (2003)
5. L. Astudillo, O. Castillo, P. Melin, A. Alanis, J. Soria, L.T. Aguilar, intelligent control of an autonomous mobile robot using type-2 fuzzy logic. Eng. Lett. **13**(3) (2006)
6. J. Urias, D. Hidalgo, P. Melin, O. Castillo, A method for response integration in modular neural networks with type-2 fuzzy logic for biometric systems. Adv. Soft Comput. **41**, 5–15 (2007)
7. R. Sepúlveda, O. Montiel, O. Castillo, P. Melin, Embedding a high speed interval type-2 fuzzy controller for a real plant into an FPGA. Appl. Soft Comput. **12**(3), 988–998 (2012)
8. O. Castillo, J.R. Castro, P. Melin, *Interval Type-3 Fuzzy Systems: Theory and Design* (Springer, Cham, Switzerland, 2022)
9. O. Castillo, P. Melin, Towards interval type-3 intuitionistic fuzzy sets and systems. Mathematics **10**, 4091 (2022)
10. M.W. Tian, S.R. Yan, A. Mohammadzadeh, J. Tavoosi, S. Mobayen, R. Safdar, W. Assawinchaichote, M.T. Vu, A. Zhilenkov, Stability of interval type-3 fuzzy controllers for autonomous vehicles. Mathematics **9**(21), 2742 (2021)
11. J.T. Rickard, J. Aisbett, G. Gibbon, Fuzzy subsethood for fuzzy sets of type-2 and generalized type-n. IEEE Trans. Fuzzy Syst. **17**(1), 50–60 (2009)
12. A. Mohammadzadeh, M.H. Sabzalian, W. Zhang, An interval type-3 fuzzy system and a new online fractional-order learning algorithm: theory and practice. IEEE Trans. Fuzzy Syst. **28**(9), 1940–1950 (2020)
13. Z. Liu, A. Mohammadzadeh, H. Turabieh, M. Mafarja, S.S. Band, A. Mosavi, A new online learned interval type-3 fuzzy control system for solar energy management systems. IEEE Access **9**, 10498–10508 (2021)
14. D. Singh, N.K. Verma, A.K. Ghosh, A.K. Malagaudanavar, An approach towards the design of interval type-3 ts fuzzy system. IEEE Trans. Fuzzy Syst. (2021)
15. J.H. Wang, J. Tavoosi, A. Mohammadzadeh, S. Mobayen, J.H. Asad, W. Assawinchaichote, P. Skruch, Non-singleton type-3 fuzzy approach for flowmeter fault detection: experimental study in a gas industry. Sensors **21**(21), 7419 (2021)
16. K.A. Alattas, A. Mohammadzadeh, S. Mobayen, A.A. Aly, B.F. Felemban, A new data-driven control system for MEMSs gyroscopes: dynamics estimation by type-3 fuzzy systems. Micromachines **12**(11), 1390 (2021)
17. Y. Cao, A. Raise, A. Mohammadzadeh, S. Rathinasamy, S.S. Band, A. Mosavi, Deep learned recurrent type-3 fuzzy system: application for renewable energy modeling/prediction. Energy Rep. **7**, 8115–8127 (2021)
18. A. Chan, J.A. Tuszynski, Automatic prediction of tumour malignancy in breast cancer with fractal dimension. R. Soc. Open Sci. **3**(12), 160558 (2016)
19. Y.D. Zhang, X.Q. Chen, T.M. Zhan, Z.Q. Jiao, Y. Sun, Z.M. Chen, S.H. Wang, Fractal dimension estimation for developing pathological brain detection system based on Minkowski-Bouligand method. IEEE Access **4**, 5937–5947 (2016)
20. O. Alpar, A mathematical fuzzy fusion framework for whole tumor segmentation in multimodal MRI using Nakagami imaging. Expert Syst. Appl. **216**, 119462 (2023)
21. O. Alpar, R. Dolezal, P. Ryska, O. Krejcar, Nakagami-fuzzy imaging framework for precise lesion segmentation in MRI. Pattern Recogn. **128**, 108675 (2022)
22. A. Di Ieva, C. Matula, F. Grizzi, G. Grabner, S. Trattnig, M. Tschabitscher, Fractal analysis of the susceptibility weighted imaging patterns in malignant brain tumors during antiangiogenic

treatment: technical report on four cases serially imaged by 7 T magnetic resonance during a period of four weeks. World Neurosurg. **77**(5–6), 785.e11–785.e21 (2012)
23. A. Di Ieva, M. Niamah, R. Menezes, M. Tsao, T. Krings, Y.B. Cho, M. Schwartz, M. Cusimano, Computational fractal-based analysis of brain arteriovenous malformation Angioarchitecture. Neurosurgery **75**(1) 72–79 (2014)
24. G. Reishofer, K. Koschutnig, C. Enzinger, F. Ebner, H. Ahammer, Fractal dimension and vessel complexity in patients with cerebral arteriovenous malformations. PLoS One **7**(7), e41148, 1–11 (2012)
25. V. Shanthoshini Deviha, P. Rengarajan, R. Jahir Hussain, Modeling blood flow in the blood vessels of the cardiovascular system using fractals. Appl. Math. Sci. **7**(11), 527–537 (2013)
26. B. González, F. Valdez, P. Melin, G. Prado-Arechiga, Fuzzy logic in the gravitational search algorithm for the optimization of modular neural networks in pattern recognition. Expert Syst. Appl. **42**(14), 5839–5847 (2015)
27. L. Amador-Angulo, O. Mendoza, J.R. Castro, A. Rodriguez-Diaz, P. Melin, O. Castillo, Fuzzy sets in dynamic adaptation of parameters of a bee colony optimization for controlling the trajectory of an autonomous mobile robot. Sensors **16**(9), 1458 (2016)
28. O. Castillo, E. Lizarraga, J. Soria, P. Melin, F. Valdez, New approach using ant colony optimization with ant set partition for fuzzy control design applied to the ball and beam system. Inf. Sci. **294**, 203–215 (2015)
29. R. Martínez-Soto, O. Castillo, L.T. Aguilar, Type-1 and Type-2 fuzzy logic controller design using a hybrid PSO–GA optimization method. Inf. Sci. **285**, 35–49 (2014)
30. A. Santiago, B. Dorronsoro, A.J. Nebro, J.J. Durillo, O. Castillo, H.J. Fraire, A novel multi-objective evolutionary algorithm with fuzzy logic based adaptive selection of operators: FAME. Inf. Sci. **471**, 233–251 (2019)

Chapter 6
A Type-3 Fuzzy-Fractal Approach for Diagnosis of Mental Disorders

In this chapter a combination of type-3 fuzzy logic and the Holder exponent (HE), which measures multi-fractality, is utilized for diagnosis of mental disorders. HE is utilized to characterize the geometrical properties of the EEG signal. A type-3 fuzzy system is employed to represent the knowledge of experts in diagnosis. In this case, type-3 helps in representing the uncertainty in making a medical diagnosis. Simulation results with a dataset of EEG signals show the effectiveness of the approach.

6.1 Introduction

Now it is known that fuzzy systems achieve striking applications in many areas, like: control, forecasting and plant monitoring. Originally, fuzzy sets (now called type-1) were put forward by Lotfi Zadeh in 1965 [1, 2]. Later fuzzy logic and fuzzy systems were also proposed by Zadeh, and many applications follow, mainly in control [3–6]. Fuzzy Logic has evolved from the original works of Zadeh with type-1 fuzzy theory [1–3], to later the developments of type-2 and its applications [7–10], to now where type-3 and its applications are emerging [11–16].

In addition, type-3 has been employed, with relative success, to handle uncertainty from sensor measurements, like in control [17–22]. There have been some previous works applying the fractal dimension (FD) to diagnosis in diverse medical problems, as in [23, 24]. Also, there exist some works using fuzzy logic in medical diagnosis, like in [25, 26]. In particular, for mental disorders a previous study has shown evidence of multi-fractality and the Holder exponents have been employed to characterize it [27]. Based on this previous work, we have extended with the use of type-3 to able to handle uncertainty. Unlike previous works, in this paper we are offering for the first time a type-3 fuzzy-fractal approach for diagnosis of mental disorders.

The rest of the document is: Sect. 6.2 is summarizing type-3 terminology, in Sect. 6.3 the fractal dimension concept is offered, in Sect. 6.4 the type-1 fuzzy fractal diagnosis is delineated, in Sect. 6.5 we are offering the type-3 fuzzy fractal diagnosis approach for mental disorders, in Sect. 6.6 results are outlined, and in Sect. 6.7 conclusions are offered.

6.2 Type-3 Fuzzy Logic

We start by postulating the concepts.

Definition 6.1 A type-3 fuzzy set (T3 FS) [22], denoted by $A^{(3)}$, is represented by the membership function (MF) of $A^{(3)}$, in the Cartesian product $X \times [0, 1] \times [0, 1]$ in $[0, 1]$, where X is the primary variable universe of $A^{(3)}$, x. The MF of $\mu_{A^{(3)}}$ is a type-3 MF (T3 MF):

$$\mu_{A^{(3)}} : X \times [0, 1] \times [0, 1] \to [0, 1]$$
$$A^{(3)} = \{(x, u(x), v(x, u), \mu_{A^{(3)}}(x, u, v)) | x \in X, u \in U \subseteq [0, 1], v \in V \subseteq [0, 1]\} \tag{6.1}$$

where U is universe for secondary variable u and V is universe for tertiary variable v. When T3MFs are employed in the rules then we have a type-3 fuzzy logic system (T3FLS).

6.3 Fractal Dimension and Holder Exponents

Recently, significant progress has been achieved in comprehending the complexity of an object by employing fractal constructs [23]. For example, time series and images in medical applications exhibit properties suggesting a fractal structure [23, 24]. The FD is postulated as:

$$d = \lim_{r \to 0} [lnN(r)]/[ln(1/r)] \tag{6.2}$$

where $N(r)$ is for number of boxes of size r. The d defined in (6.2) is approximated with box covering for several r sizes and then employing regression for finding the d value.

The degree of singularity of the signal x(t) at the point t_0 is described by the Hölder exponent, h (t_0), the largest exponent such that the analyzed signal in a neighborhood of the point t_0 can be represented as the sum of the regular component (a polynomial $P_n(t)$ of order n < h (t_0)) and the non-regular component:

6.4 Type-1 Fuzzy-Fractal Approach

$$x(t) = P_n(t) + c|t - t_0|^{h(t0)}, \quad (6.3)$$

where c is a positive constant [27]. The width of the singularity spectrum, $\Delta h = h_{min} - h_{max}$, is a measure expressing the degree of multifractality of the signal since the small Δh value reflects that the time series tends to be monofractal and the large Δh value validates the enhancement of multifractality. In this case, this Δh value is the ones that is used.

6.4 Type-1 Fuzzy-Fractal Approach

We are now presenting a fuzzy fractal approach for diagnosis of mental disorders. In Fig. 6.1 we can notice the structure of the fuzzy system in which the Holder exponent increment (HEI) of the EEG signal enters as input, and the diagnosis is the output.

In Fig. 6.2 we can find the MFs for the HEI, which are Low, Medium and High.

In Fig. 6.3 we can find the MFs for the output, which is the diagnosis (DIAG). In this case, there are three MFs, one for Normal (N), another for Depression (De) and the last one for Schizophrenia (Sc).

Table 6.1 summarizes the MF parameter values of the HEI input. The range for the HEI is from 0.5 to 1.0, as experimentally it has been found that the HEI values of the EEG signals are in this range. Table 6.2 presents the parameters for the MFs of DIAG, which we consider to be in the range from 0 to 4. All the MFs are Gaussian, so we only have two parameters, center and deviation.

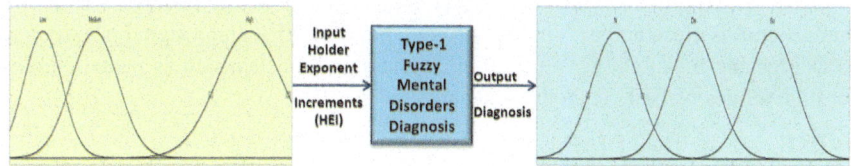

Fig. 6.1 Structure of the fuzzy fractal approach for mental disorder diagnosis

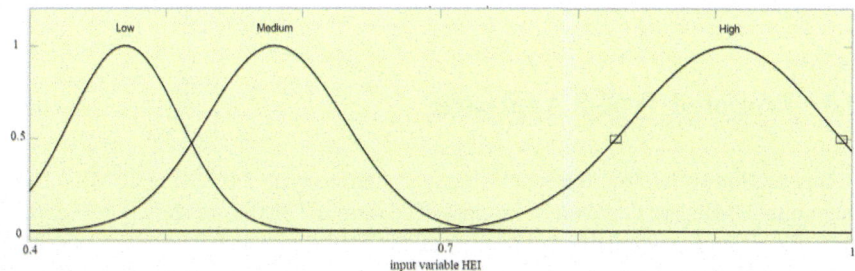

Fig. 6.2 MFs for the HEI input

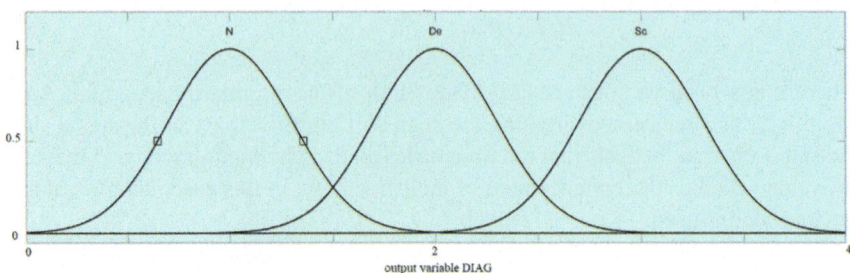

Fig. 6.3 MFs for the DIAG output variable

Table 6.1 Parameter values of the input MFs

MFs	Center	Deviation
Low	0.47	0.04
Medium	0.58	0.05
High	0.91	0.07

Table 6.2 Parameter values of the output MFs

MFs	Center	Deviation
N	1.0	0.30
De	2.0	0.30
Sc	3.0	0.30

Based on expert knowledge and experimentation with an EEG dataset it is known that a signal is classified as Normal for low values of HEI, and a signal is classified as Depression for medium values of HEI, and signal is classified as Schizophrenia for high values of HEI. Then the rules are:

If HE is Low Then DIAG is N
IF HE is Medium Then DIAG is De
If HE is High Then DIAG is Sc

We plan to employ other input variables for the system in future work.

6.5 Proposed Type-3 Approach

In this section we elevate the type-1 fuzzy system to type-3 by using T3MFs. Now the input T3MFs are depicted in Fig. 6.4. The output T3MFs are shown in Fig. 6.5.

In Fig. 6.6 we can find the architecture of the T3FLS in which the Holder exponent of the EEG signal enters as input, and the diagnosis is the output.

6.5 Proposed Type-3 Approach

Fig. 6.4 Input T3MFs

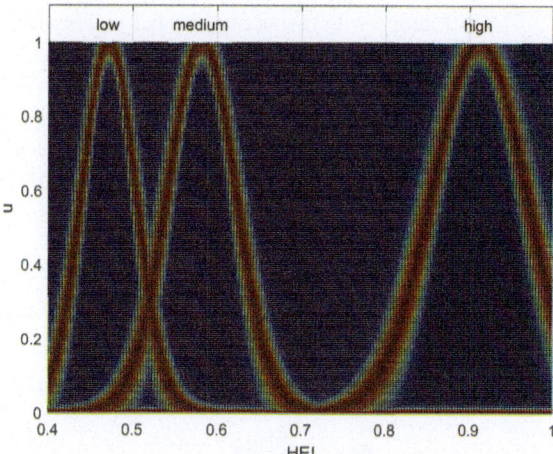

Fig. 6.5 Output T3MFs

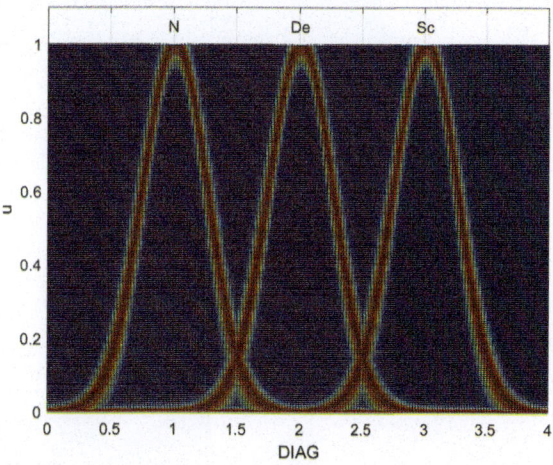

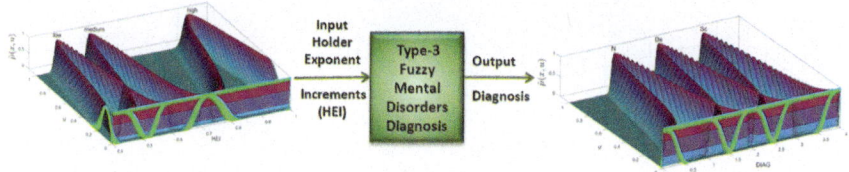

Fig. 6.6 Architecture of the T3FLS for mental disorder diagnosis

The difference with respect to Fig. 6.1 is that now we are employing T3MFs that can handle higher levels of uncertainty, which would in enhance the diagnosis. We also show in Figs. 6.7 and 6.8 a three-dimensional (3-D) view of T3MFs for the input and output, respectively.

Table 6.3 summarizes the parameter values for the T3MFs of the FD input. Table 6.4 lists the parameters for the T3MFs of DIAG, which we used in the range from 0 to 4. The T3MFs are scaled Gaussians, so we now have four parameters: center, deviation, lower lag and lower scale [22].

The rules are the same, as now the difference is that the T3FLS employs T3MFs.

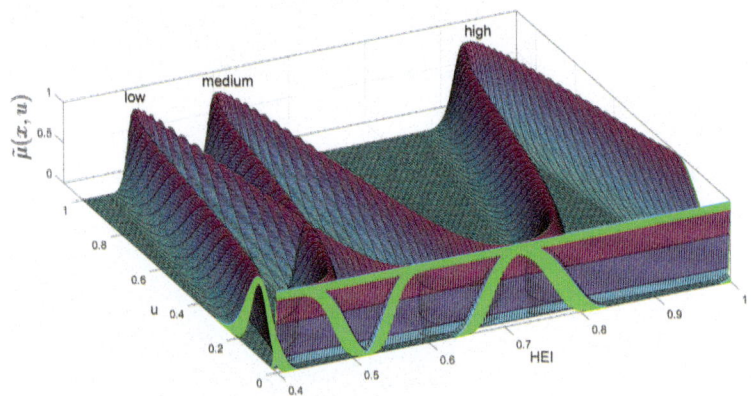

Fig. 6.7 3-D view of input T3MFs

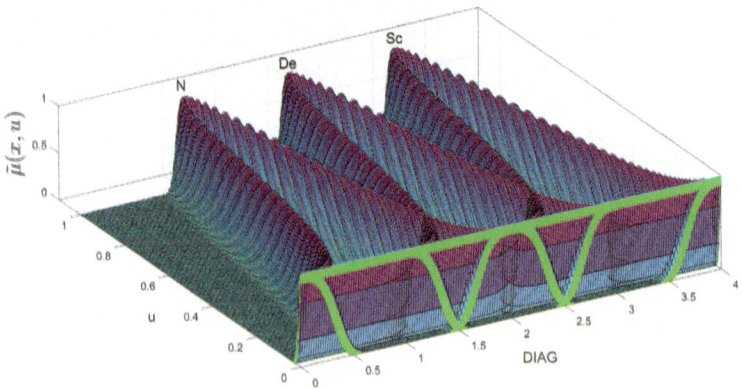

Fig. 6.8 3-D view of output T3MFs

6.6 Application to Mental Disorder Diagnosis

Table 6.3 Parameter values of the input T3MFs

MFs	Center	Deviation	Lower scale	Lower lag
Low	0.47	0.04	0.9	0.2
Medium	0.58	0.05	0.9	0.2
High	0.91	0.07	0.9	0.2

Table 6.4 Parameter values of the output T3MFs

MFs	Center	Deviation	Lower scale	Lower lag
N	1.0	0.3	0.9	0.1
De	2.0	0.3	0.9	0.1
Sc	3.0	0.3	0.9	0.1

6.6 Application to Mental Disorder Diagnosis

In this section we are presenting the results of the fuzzy-fractal approach for diagnosis of mental disorders. Table 6.5 summarizes a comparison of results, where column one shows the HEI values for 12 cases, column two lists the corresponding diagnosis of experts (N = 1, De = 2, and Sc = 3), column three shows the results of type-1, and column four the results of type-3.

From Table 6.5 we can note that both fuzzy-fractal approaches produce a correct diagnosis in 12 cases, however numerically the results of type-3 are closer on average to values of experts.

Table 6.5 Comparison of fuzzy-fractal approaches for mental disorder diagnosis

HEI	DIAG	Type-1 fuzzy DIAG	Type-3 fuzzy DIAG
0.49	N	1.25 (N)	1.1443 (N)
0.46	N	1.09 (N)	1.0393 (N)
0.51	N	1.41 (N)	1.3663 (N)
0.53	De	1.61 (De)	1.7058 (De)
0.59	De	1.98 (De)	1.9975 (De)
0.64	De	2.00 (De)	2.0037 (De)
0.57	De	1.93 (De)	1.9926 (De)
0.89	Sc	3.00 (Sc)	2.9787 (Sc)
0.92	Sc	3.00 (Sc)	2.9789 (Sc)
0.96	Sc	3.00 (Sc)	2.9752 (Sc)
0.70	De	2.15 (De)	2.0580 (De)
0.80	Sc	3.00 (Sc)	2.9233 (Sc)

6.7 Conclusions

In this chapter a hybrid of T3FLS and HEI was utilized for diagnosis of mental disorders. HEI was utilized to characterize the geometrical properties of the EEG signal. A type-3 fuzzy system was employed to represent the knowledge of experts in diagnosis. Simulation results with a dataset of EEG signals show the effectiveness of the approach. In future work we will apply the approach to the diagnosis of other medical problems. Finally, we could also consider optimizing the parameters of the fuzzy system with metaheuristics, such as the ones in [5, 28–30].

References

1. L.A. Zadeh, Fuzzy sets. Inf. Control **8**, 338–353 (1965)
2. L.A. Zadeh, Knowledge representation in fuzzy Logic. IEEE Trans. Knowl. Data Eng. **1**, 89 (1989)
3. P. Melin, O. Castillo, Adaptive intelligent control of aircraft systems with a hybrid approach combining neural networks, fuzzy logic and fractal theory. Appl. Soft Comput. **3**(4), 353–362 (2003)
4. O. Castillo, P. Melin, Intelligent adaptive model-based control of robotic dynamic systems with a hybrid fuzzy-neural approach. Appl. Soft Comput. **3**(4), 363–378 (2003)
5. L. Amador-Angulo, O. Mendoza, J.R. Castro, A. Rodriguez-Diaz, P. Melin, O. Castillo, Fuzzy sets in dynamic adaptation of parameters of a bee colony optimization for controlling the trajectory of an autonomous mobile robot. Sensors **16**(9), 1458 (2016)
6. O. Castillo, F. Kutlu, O. Atan, Intuitionistic fuzzy control of twin rotor multiple input multiple output systems. J. Intell. Fuzzy Syst. **38**(1), 821–833 (2020)
7. J.E. Moreno, M.A. Sanchez, O. Mendoza, A. Rodriguez-Diaz, O. Castillo, P. Melin, J.R. Castro, Design of an interval type-2 fuzzy model with justifiable uncertainty. Inf. Sci. **513**, 206–221 (2020)
8. L. Astudillo, O. Castillo, P. Melin, A. Alanis, J. Soria, L.T. Aguilar, Intelligent control of an autonomous mobile robot using type-2 fuzzy logic. Eng. Lett. **13**(3) (2006)
9. J. Urias, D. Hidalgo, P. Melin, O. Castillo, A method for response integration in modular neural networks with type-2 fuzzy logic for biometric systems. Adv. Soft Comput. **41**, 5–15 (2007)
10. R. Sepúlveda, O. Montiel, O. Castillo, P. Melin, Embedding a high speed interval type-2 fuzzy controller for a real plant into an FPGA. Appl. Soft Comput. **12**(3), 988–998 (2012)
11. S.N. Qasem, A. Ahmadian, A. Mohammadzadeh, S. Rathinasamy, B. Pahlevanzadeh, A type-3 logic fuzzy system: optimized by a correntropy based Kalman filter with adaptive fuzzy kernel size. Inform. Sci. **572**, 424–443 (2021)
12. A. Mohammadzadeh, M.H. Sabzalian, W. Zhang, An interval type-3 fuzzy system and a new online fractional-order learning algorithm: theory and practice. IEEE Trans. Fuzzy Syst. **28**(9), 1940–1950 (2020)
13. Z. Liu, A. Mohammadzadeh, H. Turabieh, M. Mafarja, S.S. Band, A. Mosavi, A new online learned interval type-3 fuzzy control system for solar energy management systems. IEEE Access **9**, 10498–10508 (2021)
14. D. Singh, N.K. Verma, A.K. Ghosh, A.K. Malagaudanavar, An approach towards the design of interval type-3 TS fuzzy system. IEEE Trans. Fuzzy Syst. (2021)
15. J.H. Wang, J. Tavoosi, A. Mohammadzadeh, S. Mobayen, J.H. Asad, W. Assawinchaichote, P. Skruch, Non-singleton type-3 fuzzy approach for flowmeter fault detection: experimental study in a gas industry. Sensors **21**(21), 7419 (2021)

References

16. K.A. Alattas, A. Mohammadzadeh, S. Mobayen, A.A. Aly, B.F. Felemban, A new data-driven control system for MEMSs gyroscopes: dynamics estimation by type-3 fuzzy systems. Micromachines **12**(11), 1390 (2021)
17. Y. Cao, A. Raise, A. Mohammadzadeh, S. Rathinasamy, S.S. Band, A. Mosavi, Deep learned recurrent type-3 fuzzy system: application for renewable energy modeling/prediction. Energy Rep. **7**, 8115–8127 (2021)
18. N. Nabipour, S.N. Qasem, K. Jermsittiparsert, Type-3 fuzzy voltage management in PV/hydrogen fuel cell/battery hybrid systems. Int. J. Hydrog. Energy **45**(56), 32478–33249 (2020)
19. A. Taghieh, A.A. Aly, B.F. Felemban, A. Althobaiti, A. Mohammadzadeh, A. Bartoszewicz, A hybrid predictive type-3 fuzzy control for time-delay multi-agent systems. Electronics **11**(1), 63 (2022)
20. R.H. Vafaie, A. Mohammadzadeh, M. Piran, A new type-3 fuzzy predictive controller for MEMS gyroscopes. Nonlinear Dyn. **106**(1), 381–403 (2021)
21. M.W. Tian, S.R. Yan, A. Mohammadzadeh, J. Tavoosi, S. Mobayen, R. Safdar, W. Assawinchaichote, M.T. Vu, A. Zhilenkov, Stability of interval type-3 fuzzy controllers for autonomous vehicles. Mathematics **9**(21), 2742 (2021)
22. O. Castillo, J.R. Castro, P. Melin, *Interval Type-3 Fuzzy Systems: Theory and Design* (Springer, Cham, Switzerland, 2022)
23. A. Chan, J.A. Tuszynski, Automatic prediction of tumour malignancy in breast cancer with fractal dimension. R. Soc. Open Sci. **3**(12), 160558 (2016)
24. Y.D. Zhang, X.Q. Chen, T.M. Zhan, Z.Q. Jiao, Y. Sun, Z.M. Chen, S.H. Wang, Fractal dimension estimation for developing pathological brain detection system based on Minkowski-Bouligand method. IEEE Access **4**, 5937–5947 (2016)
25. O. Alpar, A mathematical fuzzy fusion framework for whole tumor segmentation in multimodal MRI using Nakagami imaging. Expert Syst. Appl. **216**, 119462 (2023)
26. O. Alpar, R. Dolezal, P. Ryska, O. Krejcar, Nakagami-fuzzy imaging framework for precise lesion segmentation in MRI. Pattern Recogn. **128**, 108675 (2022)
27. O.E. Dick, S.V. Muraveva, V.S. Lebedev, Y.E. Shelepin, Fractal structure of brain electrical activity of patients with mental disorders. Front. Physiol. **13** (2022). https://doi.org/10.3389/fphys.2022.905318
28. B. González, F. Valdez, P. Melin, G. Prado-Arechiga, Fuzzy logic in the gravitational search algorithm for the optimization of modular neural networks in pattern recognition. Expert Syst. Appl. **42**(14), 5839–5847 (2015)
29. O. Castillo, E. Lizarraga, J. Soria, P. Melin, F. Valdez, New approach using ant colony optimization with ant set partition for fuzzy control design applied to the ball and beam system. Inf. Sci. **294**, 203–215 (2015)
30. R. Martínez-Soto, O. Castillo, L.T. Aguilar, Type-1 and Type-2 fuzzy logic controller design using a hybrid PSO–GA optimization method. Inf. Sci. **285**, 35–49 (2014)

Chapter 7
A Type-3 Fuzzy-Fractal Approach for Diagnosis of Vascular Diseases Based on Cerebral Vessels

In this chapter a combination of type-3 fuzzy logic system (T3FLS) and the fractal dimension (FD) is employed for diagnosis of vascular cerebral diseases. FD is utilized to characterize the geometrical properties of the cerebral vessels. A T3FLS is employed to represent the knowledge of experts in vascular cerebral diagnosis. In this situation, the T3FLS helps in representing the uncertainty in making a medical diagnosis. Simulation results with a dataset of x-ray images of cerebral vessels shows the accuracy of the proposal.

7.1 Introduction

Nowadays it is accepted that fuzzy systems attain remarkable applications in many areas, like: control, forecasting and monitoring. Originally, fuzzy sets offered by Zadeh in 1965 [1, 2]. Later fuzzy systems were also offered by Zadeh, and many applications follow, like in [3, 4]. Fuzzy Logic has evolved from the initial studies of Zadeh with type-1 fuzzy theory [1], to later the theory of type-2 [5–7], to now where type-3 is appearing [8–10]. Now even type-n has been mentioned [11]. In addition, type-3 has been employed to handle uncertainty from measurements, like in control [12–17]. There have been some previous works applying the fractal dimension (FD) to diagnosis in diverse medical problems, as in [18, 19]. Also, there exist some works using fuzzy logic in medical diagnosis, like in [20, 21].

The vascular disease that is considered in this chapter is the arterio-venous malformation (AVM) which appears at the connection of the artery and vein. Normally, this connection is made with the aid of the capillaries which are the smallest vessels from the circulatory system and they have the role of feeding the cells. For this disease, the capillaries are missing and the exchanging of the oxygenated blood with the cells is impaired. There are previous works on utilizing fractal methods for this problem, like in [22–25]. Unlike previous works, in this paper we are offering for the first time a type-3 fuzzy-fractal hybrid for diagnosis of vascular diseases based on cerebral vessels. We employ values from a dataset of x-ray images from cerebral vessels [26] for testing our approach.

The rest of the document is: Sect. 7.2 is summarizing type-3 terminology, in Sect. 7.3 the fractal dimension concept is offered, in Sect. 7.4 the type-1 fuzzy fractal diagnosis is delineated, in Sect. 7.5 we are offering the type-3 fuzzy fractal vascular cerebral diagnosis approach, in Sect. 7.6 results are presented, and in Sect. 7.7 conclusions are offered.

7.2 Type-3 Fuzzy Theory

We first postulate the concepts.

Definition 7.1 A type-3 fuzzy set (T3 FS) [8], written as $A^{(3)}$, is represented by the membership function (MF) of $A^{(3)}$, in the Cartesian product $X \times [0, 1] \times [0, 1]$ in $[0, 1]$, where X is the primary variable universe of $A^{(3)}$, x. The MF of $\mu_{A^{(3)}}$ is a type-3 MF (T3 MF):

$$\mu_{A^{(3)}} : X \times [0, 1] \times [0, 1] \to [0, 1]$$
$$A^{(3)} = \{(x, u(x), v(x, u), \mu_{A^{(3)}}(x, u, v)) | x \in X, u \in U \subseteq [0, 1], v \in V \subseteq [0, 1]\} \tag{7.1}$$

where U is universe for secondary variable u and V is universe for tertiary variable v. When T3MFs are employed in the rules then we have a type-3 fuzzy logic system (T3FLS).

7.3 Fractal Dimension

Recently, remarkable advances have been achieved in understanding the complexity of an object by employing fractal theory [18, 19]. For example, time series in finance suggest fractal structure. There are several definitions of FD, but here we utilize one of the simplest ones. The FD is postulated as:

$$d = \lim_{r \to 0} [lnN(r)]/[ln(1/r)] \tag{7.2}$$

where $N(r)$ is for number of boxes of size r. The d defined in (7.2) is approximated with box covering for several r sizes and then employing regression for estimating the d value.

The study was done using cerebral angiograms. This clinical procedure is the gold standard for diagnosis of the arterio-venous malformations [26]. In the acquired data, the AVM is located into the images as a dashed irregular shape for two different image series. The medical expert selects this location and then a region with healthy vessels in order to compare their FDs. As in other cases, some image processing steps (like edge detection) are undergone and the disease has a higher FD than a normal case [26]. The FD augmentation is related to structural vascular complexity due to the increased number of feeding arteries. For example, the FD value for an image with an AVM case is about 1.937, while for a normal case is about 1.915. This kind of a difference in FD values is consistent and for this reason can be used as a basis for the decision-making process required in diagnosing the cerebral disease.

7.4 Type-1 Fuzzy-Fractal Approach

We are presenting a fuzzy fractal approach for diagnosis of vascular cerebral diseases. In Fig. 7.1 we can notice the fuzzy system structure, in which the fractal dimension (FD) of the cerebral vessels [26] enters as input, and the diagnosis is the output.

In Fig. 7.2 we can find the MFs for the FD, which are Low and High. In Fig. 7.3 we can find the MFs for the output, which is the diagnosis (DIAG). In this case, there are two MFs, one for Healthy (He) and the other for AVM (A), which is an arteriovenous malformation is *a tangle of blood vessels that irregularly connects arteries and veins.*

Table 7.1 summarizes the MF parameters for the FD input. The range for the FD is from 1.90 to 1.94, as experimentally it has been found that the FDs of the cerebeal vessels are in this range. Table 7.2 presents the parameters for the MFs of DIAG,

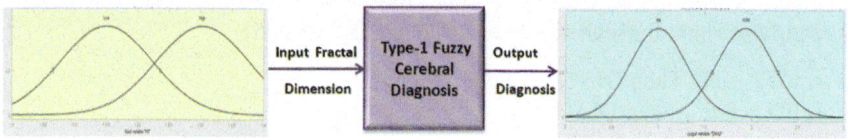

Fig. 7.1 Structure of the fuzzy fractal approach for cerebral vascular diagnosis

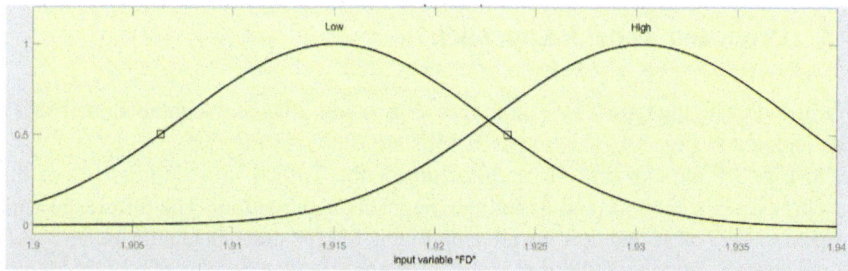

Fig. 7.2 MFs for the FD input

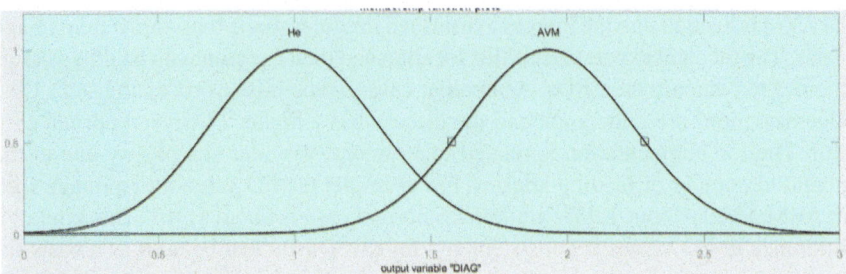

Fig. 7.3 MFs for the DIAG output variable

Table 7.1 Parameter values of the input MFs	MFs	Center	Deviation
	Low	1.915	0.0073
	High	1.930	0.0073

Table 7.2 Parameter values of the output MFs	MFs	Center	Deviation
	He	1.0	0.30
	St	2.0	0.30

which we consider to be in the range from 0 to 3. All MFs are Gaussian, so we have 2 parameters, center and deviation.

Based on expert knowledge and experimentation with the dataset it is known that an image is classified as Healthy (normal) for low values of FD, and is classified as AVM (dangerous) for high values of FD. Then the rules are:

If FD is Low Then DIAG is He
If FD is High Then DIAG is A

We think of employing other input variables for the fuzzy system in future work.

7.5 Proposed Type-3 Approach

We now elevate the type-1 system to type-3 by using T3MFs. Now the input T3MFs are depicted in Fig. 7.4. The output T3MFs are shown in Fig. 7.5.

In Fig. 7.6 we can find the architecture of the T3FLS in which the FD of the cerebral vessels enters as input, and the diagnosis is the output. The difference with respect to Fig. 7.1 is that now we are employing T3MFs that can handle higher levels of uncertainty, which would in enhance the diagnosis. We also show in Figs. 7.7 and 7.8 a three-dimensional (3-D) view of T3MFs for the input and output, respectively.

7.5 Proposed Type-3 Approach

Fig. 7.4 Input T3MFs

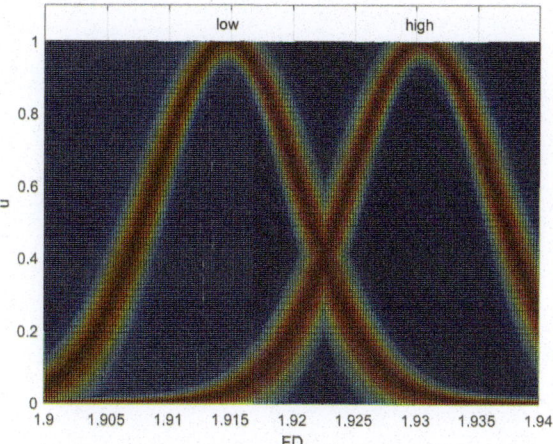

Fig. 7.5 Output T3MFs

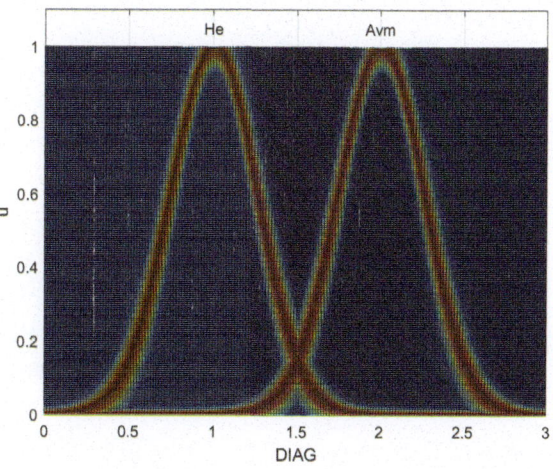

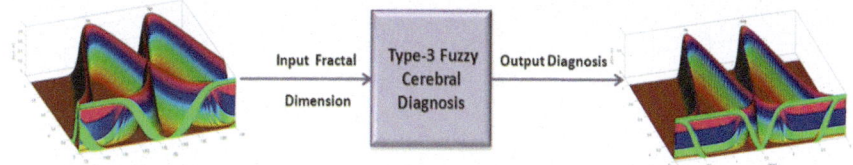

Fig. 7.6 Architecture of the T3FLS for diagnosis of cerebral vascular diseases

Table 7.3 summarizes the parameter values for the T3MFs of the FD input. Table 7.4 lists the parameters for the T3MFs of DIAG, which we used in the range from 0 to 3. The T3MFs are scaled Gaussians, so we now have four parameters: center, deviation, lower lag and lower scale [8].

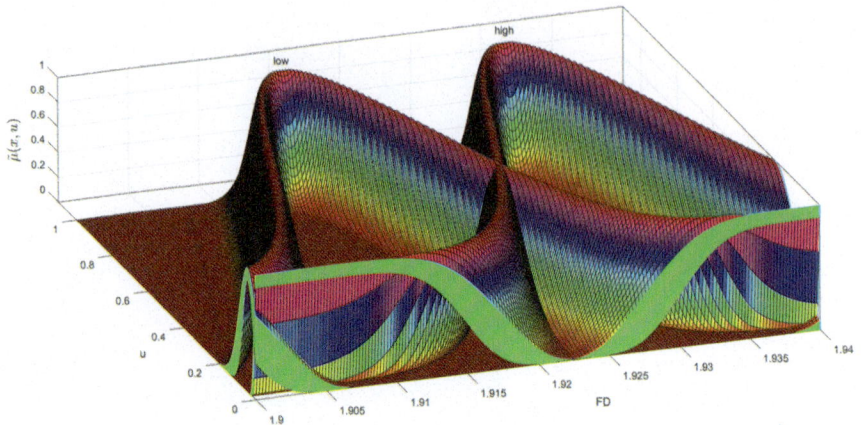

Fig. 7.7 3-D view of input T3MFs

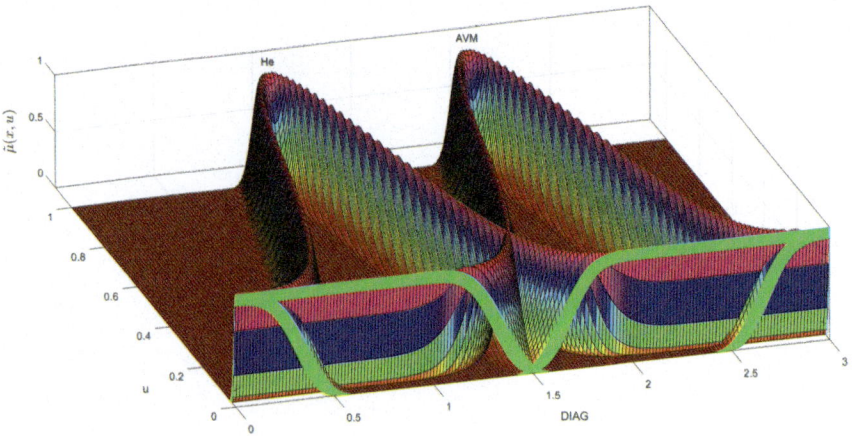

Fig. 7.8 3-D view of output T3MFs

Table 7.3 Parameter values of the input T3MFs

MFs	Center	Deviation	Lower scale	Lower lag
Low	1.9146	0.0073	0.9	0.2
High	1.9303	0.0073	0.9	0.2

Table 7.4 Parameter values of the output T3MFs

MFs	Center	Deviation	Lower scale	Lower lag
He	1.0	0.3	0.9	0.1
A	2.0	0.3	0.9	0.1

7.7 Conclusions

Table 7.5 Comparison of fuzzy-fractal approaches for diagnosis of vascular diseases

FD	DIAG	Type-1 fuzzy DIAG	Type-3 fuzzy DIAG
1.9050	He	1.01 (He)	1.0216 (He)
1.9100	He	1.04 (He)	1.0136 (He)
1.9150	He	1.15 (He)	1.0581 (He)
1.9200	He	1.35 (He)	1.2942 (He)
1.9250	A	1.57 (A)	1.7137 (A)
1.9275	A	1.68 (A)	1.8627 (A)
1.9300	A	1.77 (A)	1.9457 (A)
1.9325	A	1.85 (A)	1.9795 (A)
1.9350	A	1.89 (A)	1.9865 (A)
1.9375	A	1.91 (A)	1.9854 (A)

The rules are the same, as now the difference is that the T3FLS employs T3MFs.

7.6 Application to Diagnosis of Cerebral Vascular Diseases

This section is presenting the results of the fuzzy-fractal systems for diagnosis of cerebral vascular diagnosis. Table 7.5 summarizes a comparison of results, where column one shows the FD values for 12 cases, column two lists the corresponding diagnosis of experts (He = 1, A = 2), column three shows the results of type-1, and column four results of type-3. From Table 7.5 we can note that both fuzzy-fractal approaches produce a correct diagnosis in 10 cases, however numerically the results of type-3 are closer on average to values of experts. In fact, in 9 of 10 cases the type-3 approach provides a numeric value that is closer to the actual value of the experts. Only, in the first case (first row) the results of type-1 are slightly better than type-3. So, we can conclude that type-3 surpasses type-1 in handling the uncertainty in the diagnosis process.

7.7 Conclusions

In this chapter a hybrid of type-3 and FD was employed for the diagnosis of cerebral vascular diseases. FD was utilized to characterize the geometrical properties of the cerebral vessels. A type-3 system was employed to represent the knowledge of experts in diagnosis. Simulation results with a dataset of x-ray images shows the effectiveness of the proposal. In future work we could employ the diagnosis approach for other medical problems. In addition, we could employ metaheuristic optimization algorithms in finding the optimal parameterization of the fuzzy systems, as has been done in works such as [27–31].

References

1. L.A. Zadeh, Fuzzy sets. Inf. Control **8**, 338–353 (1965)
2. L.A. Zadeh, Knowledge representation in fuzzy logic. IEEE Trans. Knowl. Data Eng. **1**, 89 (1989)
3. P. Melin, O. Castillo, Adaptive intelligent control of aircraft systems with a hybrid approach combining neural networks, fuzzy logic and fractal theory. Appl. Soft Comput. **3**(4), 353–362 (2003)
4. O. Castillo, P. Melin, Intelligent adaptive model-based control of robotic dynamic systems with a hybrid fuzzy-neural approach. Appl. Soft Comput. **3**(4), 363–378 (2003)
5. L. Astudillo, O. Castillo, P. Melin, A. Alanis, J. Soria, L.T. Aguilar, Intelligent control of an autonomous mobile robot using type-2 fuzzy logic. Eng. Lett. **13**(3) (2006)
6. J. Urias, D. Hidalgo, P. Melin, O. Castillo, A method for response integration in modular neural networks with type-2 fuzzy logic for biometric systems. Adv. Soft Computing **41**, 5–15 (2007)
7. R. Sepúlveda, O. Montiel, O. Castillo, P. Melin, Embedding a high speed interval type-2 fuzzy controller for a real plant into an FPGA. Appl. Soft Comput. **12**(3), 988–998 (2012)
8. O. Castillo, J.R. Castro, P. Melin, *Interval Type-3 Fuzzy Systems: Theory and Design* (Springer, Cham, Switzerland, 2022)
9. O. Castillo, P. Melin, Towards interval type-3 intuitionistic fuzzy sets and systems. Mathematics **10**, 4091 (2022)
10. M.W. Tian, S.R. Yan, A. Mohammadzadeh, J. Tavoosi, S. Mobayen, R. Safdar, W. Assawinchaichote, M.T. Vu, A. Zhilenkov, Stability of interval type-3 fuzzy controllers for autonomous vehicles. Mathematics **9**(21), 2742 (2021)
11. J.T. Rickard, J. Aisbett, G. Gibbon, Fuzzy subsethood for fuzzy sets of type-2 and generalized type-n. IEEE Trans. Fuzzy Syst. **17**(1), 50–60 (2009)
12. A. Mohammadzadeh, M.H. Sabzalian, W. Zhang, An interval type-3 fuzzy system and a new online fractional-order learning algorithm: theory and practice. IEEE Trans. Fuzzy Syst. **28**(9), 1940–1950 (2020)
13. Z. Liu, A. Mohammadzadeh, H. Turabieh, M. Mafarja, S.S. Band, A. Mosavi, A new online learned interval type-3 fuzzy control system for solar energy management systems. IEEE Access **9**, 10498–10508 (2021)
14. D. Singh, N.K. Verma, A.K. Ghosh, A.K. Malagaudanavar, An approach towards the design of interval type-3 TS fuzzy system. IEEE Trans. Fuzzy Syst. (2021)
15. J.H. Wang, J. Tavoosi, A. Mohammadzadeh, S. Mobayen, J.H. Asad, W. Assawinchaichote, P. Skruch, Non-singleton type-3 fuzzy approach for flowmeter fault detection: experimental study in a gas industry. Sensors **21**(21), 7419 (2021)
16. K.A. Alattas, A. Mohammadzadeh, S. Mobayen, A.A. Aly, B.F. Felemban, A new data-driven control system for MEMSs gyroscopes: dynamics estimation by type-3 fuzzy systems. Micromachines **12**(11), 1390 (2021)
17. Y. Cao, A. Raise, A. Mohammadzadeh, S. Rathinasamy, S.S. Band, A. Mosavi, Deep learned recurrent type-3 fuzzy system: application for renewable energy modeling/prediction. Energy Rep. **7**, 8115–8127 (2021)
18. A. Chan, J.A. Tuszynski, Automatic prediction of tumour malignancy in breast cancer with fractal dimension. R. Soc. Open Sci. **3**(12), 160558 (2016)
19. Y.D. Zhang, X.Q. Chen, T.M. Zhan, Z.Q. Jiao, Y. Sun, Z.M. Chen, S.H. Wang, Fractal dimension estimation for developing pathological brain detection system based on Minkowski-Bouligand method. IEEE Access **4**, 5937–5947 (2016)
20. O. Alpar, A mathematical fuzzy fusion framework for whole tumor segmentation in multimodal MRI using Nakagami imaging. Expert Syst. Appl. **216**, 119462 (2023)
21. O. Alpar, R. Dolezal, P. Ryska, O. Krejcar, Nakagami-fuzzy imaging framework for precise lesion segmentation in MRI. Pattern Recogn. **128**, 108675 (2022)
22. A. Di Ieva, C. Matula, F. Grizzi, G. Grabner, S. Trattnig, M. Tschabitscher, Fractal analysis of the susceptibility weighted imaging patterns in malignant brain tumors during antiangiogenic

treatment: technical report on four cases serially imaged by 7 T magnetic resonance during a period of four weeks. World Neurosurg. **77**(5–6), 785.e11–785.e21 (2012)
23. A. Di Ieva, M. Niamah, R. Menezes, M. Tsao, T. Krings, Y.B. Cho, M. Schwartz, M. Cusimano, Computational fractal-based analysis of brain arteriovenous malformation angioarchitecture. Neurosurgery **75**(1), 72–79 (2014)
24. G. Reishofer, K. Koschutnig, C. Enzinger, F. Ebner, H. Ahammer, Fractal dimension and vessel complexity in patients with cerebral arteriovenous malformations. PLoS One **7**(7), e41148, 1–11(2012)
25. V. Shanthoshini Deviha, P. Rengarajan, R. Jahir Hussain, Modeling blood flow in the blood vessels of the cardiovascular system using fractals. Appl. Math. Sci. **7**(11), 527–537 (2013)
26. T.I. Andra, D. Popescu, C. Vasseur, Fractal dimension for the analysis of vascular diseases on x-ray angiography, in *20th International Conference on Control Systems and Computer Science* (Bucarest, Romania, 2015). https://doi.org/10.1109/CSCS.2015.88
27. B. González, F. Valdez, P. Melin, G. Prado-Arechiga, Fuzzy logic in the gravitational search algorithm for the optimization of modular neural networks in pattern recognition. Expert Syst. Appl. **42**(14), 5839–5847 (2015)
28. L. Amador-Angulo, O. Mendoza, J.R. Castro, A. Rodriguez-Diaz, P. Melin, O. Castillo, Fuzzy sets in dynamic adaptation of parameters of a bee colony optimization for controlling the trajectory of an autonomous mobile robot. Sensors **16**(9), 1458 (2016)
29. O. Castillo, E. Lizarraga, J. Soria, P. Melin, F. Valdez, New approach using ant colony optimization with ant set partition for fuzzy control design applied to the ball and beam system. Inf. Sci. **294**, 203–215 (2015)
30. R. Martínez-Soto, O. Castillo, L.T. Aguilar, Type-1 and type-2 fuzzy logic controller design using a hybrid PSO–GA optimization method. Inf. Sci. **285**, 35–49 (2014)
31. A. Santiago, B. Dorronsoro, A.J. Nebro, J.J. Durillo, O. Castillo, H.J. Fraire, A novel multi-objective evolutionary algorithm with fuzzy logic based adaptive selection of operators: FAME. Inf. Sci. **471**, 233–251 (2019)

Chapter 8
Conclusions of Type-3 Fuzzy-Fractal Medical Diagnosis

The basic terminology and methods for designing type-3 fuzzy sets, membership functions, inference and fuzzy systems for medical diagnosis have been outlined in this book [1]. In addition, fractal theory (FT) concepts, such as the fractal dimension (FD) and Hurst exponents, have been outlined and applied to medical diagnosis. So, for the case of designing type-3 fuzzy-fractal systems, the problem is more complicated due to the higher number of parameters to consider, and we have involved the utilization of nature-inspired optimization techniques for this problem [2].

The main idea of the work has been to introduce in a systematic way the concepts and methods of interval type-3 fuzzy systems that with their higher uncertainty handling capabilities we expect that will solve, in a better way, more difficult medical diagnosis problems [3, 4].

This monograph has offered several interesting applications in medical diagnosis to illustrate the potential of combining type-3 fuzzy logic (T3FL) with fractal theory. Chapter 2 deals with tuberculosis diagnosis based on the FD and T3FL. Chapter 3 outlines the type-3 fuzzy-fractal approach for bone analysis and osteoporosis diagnosis. Chapter 4 deals with Leukemia cancer diagnosis based on type-3 and FT. Chapter 5 describes the diagnosis for vascular diseases in cardiac vessels. Chapter 6 delineates an approach for diagnosis of mental disorders combining type-3 and FT. Chapter 7 describes the diagnosis for vascular diseases in cerebral vessels. Finally, Chapter 8 outlines the conclusions.

As future works, we consider that it will be interesting to build type-3 fuzzy systems for specific problems in different areas of application, such as intelligent control [5, 6], robotics [7, 8], pattern recognition [9], optimization [10, 11], diagnosis in different cases [12] and other problems [13–15].

There are also nature-inspired algorithms that have not yet been applied in designing type-3, like for example: plant self-defense algorithm, electromagnetism-based algorithm, and similar ones [16, 17]. In addition, another research area has been the utilization of fuzzy systems for achieving adaptation in metaheuristics, and we envision a direct utilization of type-3 in this area, as is presented in [18–20].

Finally, it is worthwhile to mention other possible avenues of research, which would be the hybrid combination of T3FL with Z numbers, also a possible hybrid type-3 neutrosophic approach could be envisioned, as well as considering type-3 Pythagorean fuzzy systems or a type-3 hesitant combination could be think of, and so on other similar interesting proposals [21].

References

1. O. Castillo, J.R. Castro, P. Melin, *Interval Type-3 Fuzzy Systems: Theory and Design* (Springer, Cham, Switzerland, 2022)
2. O. Castillo, P. Melin, Towards interval type-3 intuitionistic fuzzy sets and systems. Mathematics **10**, 4091 (2022). https://doi.org/10.3390/math10214091
3. Z. Liu, A. Mohammadzadeh, H. Turabieh, M. Mafarja, S.S. Band, A. Mosavi, A new online learned interval type-3 fuzzy control system for solar energy management systems. IEEE Access **9**, 10498–10508 (2021)
4. R.H. Vafaie, A. Mohammadzadeh, M. Piran, A new type-3 fuzzy predictive controller for MEMS gyroscopes. Nonlinear Dyn. **106**(1), 381–403 (2021)
5. L. Aguilar, P. Melin, O. Castillo, Intelligent control of a stepping motor drive using a hybrid neuro-fuzzy ANFIS approach. Appl. Soft Comput. **3**(3), 209–219
6. P. Melin, O. Castillo, Adaptive intelligent control of aircraft systems with a hybrid approach combining neural networks, fuzzy logic and fractal theory. Appl. Soft Comput. **3**(4), 353–362 (2003)
7. O. Castillo, P. Melin, A new fuzzy-fractal-genetic method for automated mathematical modelling and simulation of robotic dynamic systems, in *1998 IEEE International Conference on Fuzzy Systems (FUZZ-IEEE 1998) Proceedings*, vol 2, pp. 1182–1187
8. O. Castillo, P. Melin, Intelligent adaptive model-based control of robotic dynamic systems with a hybrid fuzzy-neural approach. Appl. Soft Comput. **3**(4), 363–378 (2003)
9. D. Sanchez, P. Melin, O. Castillo, A grey wolf optimizer for modular granular neural networks for human recognition. Comput. Intell. Neurosci. **2017**, 4180510:1–4180510:26 (2017)
10. F. Valdez, P. Melin, O. Castillo, Evolutionary method combining particle swarm optimization and genetic algorithms using fuzzy logic for decision making, in: *IEEE International Conference on Fuzzy Systems*, pp. 2114–2119 (2009)
11. F. Valdez, J.C. Vazquez, P. Melin, O. Castillo, Comparative study of the use of fuzzy logic in improving particle swarm optimization variants for mathematical functions using co-evolution. Appl. Soft Comput. **52**, 1070–1083 (2017)
12. E. Ontiveros, P. Melin, O. Castillo, Comparative study of interval type-2 and general type-2 fuzzy systems in medical diagnosis. Inf. Sci. **525**, 37–53 (2020)
13. H.I. Seker, S. Kacar, O. Castillo, S. Uzun, I. Pehlivan, Z. Tatli, Detection of resistance spot welding faults in copper materials by transfer learning method. Appl. Comput. Math. **22**(3), 430–445 (2023). https://doi.org/10.30546/1683-6154.22.3.2023.430
14. D. Mohapatra, S. Chakraverty, O. Castillo, Numerical investigation of fluid dynamic model in uncertain environment. Appl. Comput. Math. **22**(3), 297–316 (2023). https://doi.org/10.30546/1683-6154.22.3.2023.297
15. F. Valdez, O. Castillo, P. Cortes-Antonio, P. Melin, Applications of intelligent optimization algorithms and fuzzy logic systems in aerospace: a review. Appl. Comput. Math. **21**(3), 233–245 (2022). https://doi.org/10.30546/1683-6154.21.3.2022.2330
16. F. Valdez, H. Carreon-Ortiz, O. Castillo, CMOA—Continuous mycorrhiza optimization algorithm, in *Mycorrhiza Optimization Algorithm*. Springer Briefs in Applied Sciences and Technology (Springer, Cham, 2023). https://doi.org/10.1007/978-3-031-47369-2_5

References

17. F. Valdez, H. Carreon-Ortiz, O. Castillo, DMOA—Discrete mycorrhiza optimization algorithm, in *Mycorrhiza Optimization Algorithm*. Springer Briefs in Applied Sciences and Technology (Springer, Cham, 2023). https://doi.org/10.1007/978-3-031-47369-2_6
18. L. Amador-Angulo, O. Castillo, P. Melin, J.R. Castro, Interval type-3 fuzzy adaptation of the bee colony optimization algorithm for optimal fuzzy control of an autonomous mobile robot. Micromachines **13**, 1490 (2022). https://doi.org/10.3390/mi13091490
19. L. Amador-Angulo, O. Castillo, J.R. Castro et al., A new approach for interval type-3 fuzzy control of nonlinear plants. Int. J. Fuzzy Syst. **25**, 1624–1642 (2023). https://doi.org/10.1007/s40815-023-01470-9
20. C. Peraza, P. Ochoa, O. Castillo, Z.W. Geem, Interval-type 3 fuzzy differential evolution for designing an interval-type 3 fuzzy controller of a unicycle mobile robot. Mathematics **10**, 3533 (2022). https://doi.org/10.3390/math10193533
21. O. Castillo, P. Melin, Proposal for mediative fuzzy control: from type-1 to type-3. Symmetry **2023**, 15 (1941). https://doi.org/10.3390/sym15101941

Index

A
Anisotropy, 15, 17, 19

B
Bone analysis, 2, 67

C
Cancer diagnosis, 2, 27, 28, 67
Cardiac vessels, 2, 37–40, 43, 67
Cerebral vessels, 2, 57–60, 63, 67
Coronary artery, 37

F
Fractal dimension, 2, 5, 6, 15, 17, 19, 27, 28, 37–39, 47, 48, 57–59, 67
Fuzzy fractal, 5–7, 15, 17, 27–29, 38, 39, 48, 49, 58, 59
Fuzzy logic, 1, 5, 6, 15, 16, 27, 28, 37, 38, 47, 48, 57, 58, 67
Fuzzy models, 1
Fuzzy rules, 6, 16, 28, 38
Fuzzy sets, 1, 5, 15, 27, 37, 47, 57, 67

I
Interval type-3 fuzzy models, 1

Interval type-3 fuzzy systems, 67

L
Leukemia, 2, 27–31, 33, 67

M
Medical diagnosis, 1, 2
Membership function, 1, 6, 16, 28, 38, 48, 58, 67
Mental disorders, 2, 47–49, 53, 54, 67

O
Osteoporosis, 2, 15–18, 67

R
Roughness, 17, 19

S
Stenosis, 37, 39

T
Tuberculosis, 2, 5, 6, 8, 11, 12, 67
Type-1 fuzzy fractal, 5, 15, 27, 38, 48, 58

Type-1 fuzzy sets, 1
Type-1 fuzzy system, 8, 40, 50
Type-1 fuzzy theory, 5, 15, 27, 37, 47, 57
Type-2 fuzzy sets, 1
Type-3 fuzzy, 1, 2, 5, 6, 11, 15, 16, 27, 28, 33, 37, 38, 47, 48, 54, 57, 58, 67

U
Uncertainty, 1, 5, 8, 15, 19, 27, 28, 31, 37, 40, 47, 52, 57, 60, 63, 67

V
Vascular diseases, 2, 37–39, 43, 58, 61, 63, 67

The manufacturer's authorised representative in the EU is Springer Nature Customer Service Centre GmbH, Europaplatz 3, 69115 Heidelberg, Germany. If you have any concerns regarding our products, please contact ProductSafety@springernature.com

Printed and bound by CPI Group (UK) Ltd, Croydon, CR0 4YY
26/03/2026
02078940-0006